# Fire in the United States
# 2003-2007

## Fifteenth Edition

I0488933

## FA-325 / October 2009

## U.S. Fire Administration

## Mission Statement

*We provide National leadership to foster a solid foundation for our fire and emergency services stakeholders in prevention, preparedness, and response.*

# Table of Contents

# List of Figures

# List of Tables

# Acknowledgements

The U.S. Fire Administration (USFA) greatly appreciates the participation in the National Fire Incident Reporting System (NFIRS) from fire departments across the United States. The NFIRS data, on which the bulk of this report is based, are available through the work of the staffs of the various State agencies and State Fire Marshals' offices responsible for fire data collection, and on each and every fire officer who fills out an NFIRS form. Without their efforts to collect data, this report could not exist. Although reporting to NFIRS is wholly voluntary, the information collected on fires each year represents the most comprehensive set of fire data and statistics in the world. At the time of publication, 20,600 fire departments submitted fire incident data; these departments added 1.3 million fire incident records to the NFIRS database in the 2007 data files alone.

The National Fire Information Council (NFIC), a nonprofit organization of State and metropolitan area participants in NFIRS, helps coordinate and specify requirements for NFIRS and its operation. NFIC represents an outstanding example of local, State, and Federal cooperation on a major, long-term undertaking. The USFA appreciates the support NFIC has provided to NFIRS over the years.

The USFA also thanks the many State Fire Marshals' offices or their equivalents for the help and assistance they provide to USFA staff.

The report was produced by TriData, a division of System Planning Corporation, Arlington, Virginia, for the National Fire Data Center, USFA, under contract HSFEEM-06A-0029, task order number HSFEEM-09-J-0001.

Copies of this report are available by writing:

U.S. Fire Administration
Federal Emergency Management Agency
Publications Center
16825 South Seton Avenue
Emmitsburg, Maryland 21727

Documents may also be ordered on the Internet at: http://www.usfa.dhs.gov/applications/publications

# Executive Summary

Fire departments in the United States responded to nearly 1.6 million fire calls in 2007. The United States fire problem, on a per capita basis, is one of the worst in the industrial world. Thousands of Americans die each year, tens of thousands of people are injured, and property losses reach billions of dollars. There are huge indirect costs of fire as well—temporary lodging, lost business, medical expenses, psychological damage, and others. These indirect costs may be as much as 8- to 10-times higher than the direct costs of fire. To put this in context, the annual losses from floods, hurricanes, tornadoes, earthquakes, and other natural disasters combined in the United States average just a fraction of those from fires. The public, the media, and local governments generally are unaware of the magnitude and seriousness of the fire problem to individuals and their families, to communities, and to the Nation.

## PURPOSE AND SCOPE

The National Fire Data Center (NFDC) of the U.S. Fire Administration (USFA) periodically publishes *Fire in the United States*, a statistical overview of the fires in the United States with the focus on the latest year in which data were available at the time of preparation. This report is designed to equip the fire service and others with information that motivates corrective action, sets priorities, targets specific fire programs, serves as a model for State and local analyses of fire data, and provides a baseline for evaluating programs.

This Fifteenth Edition covers the 5-year period of 2003 to 2007 with a primary focus on 2007. Only native National Fire Incident Reporting System (NFIRS) 5.0 data are used for NFIRS-based analyses. In 2007, the native NFIRS 5.0 data account for 98 percent of the fire incident data.

The report addresses the overall national fire problem. Detailed analyses of the residential and non-residential fire problem, firefighter casualties, and other subsets of the national fire problem are not included. These topic-specific analyses are addressed as separate, stand-alone publications.

The primary source of data is from NFIRS. National Fire Protection Association (NFPA) annual survey results, mortality data from the National Center for Health Statistics (NCHS), data from State Fire Marshals' offices or their equivalents, population data from the U.S. Census Bureau, and inflation adjustments from the Bureau of Labor Statistic's Consumer Price Index also are used. Because of the time it takes for States to submit data to USFA from the thousands of fire departments that participate in NFIRS, then obtain corrections and edit the data, and analyze and display the results, the publication lags behind the date of data collection. Fortunately, the fire problem does not change very rapidly, so the data usually are quite representative of the situation in the year of publication as well.

1

# NATIONAL PROBLEM

Annual deaths from fire in the United States were estimated at 12,000 in 1974, the year in which the USFA was established. At that time, a goal was set for reducing this number by half within a generation. This goal was met. By 2006, NFPA estimates of civilian deaths were at their lowest level (3,245). While fire deaths are still trending downward, in 2007, NFPA estimates of fire deaths were 6 percent higher than they were in 2006.[1]

Table 1 presents 5-year fire and fire loss rate trends. Fires per million population reached a new low in 2007 and continue a downward trend. Property loss, while up substantially from 2006, increased just 4 percent over the 5 years. Injuries and deaths per million population continue to decline despite the 1-year increase in 2007. The death rate (deaths per million population) declined nearly 20 percent between 2003 and 2007 and is less than half of what it was in the late 1970s.[2] Nevertheless, the United States has a fire death rate 2 to 2 1/2 times that of several European Nations and at least 20 percent higher than many other Nations. Of the 25 industrial Nations examined by the World Fire Statistics Centre, the United States ranked as having the fifth highest fire death rate. This general status has been unchanged for the past 27 years.

### Table 1. Fire and Fire Loss Rate Trends

| Loss Measure | 5-Year Trend (percent) |
|---|---|
| Fires/Million Population | -2.7 |
| Deaths/Million Population | -19.7 |
| Injuries/Million Population | -8.8 |
| Dollar Loss/Capita* | 3.8 |

*Trend adjusted to 2007 dollars

Sources:  NFPA, Consumer Price Index, and U.S. Census Bureau.

# REGIONAL AND STATE PROFILES

The fire problem varies from region to region and State to State because of variations in climate, socio-economic status, education, demographics, and other factors. Five States (Alaska, Mississippi, Kansas, Tennessee, and West Virginia) and the District of Columbia have fire death rates that exceed 25 deaths per million population; this rate is one of the worst among the world's Nations. Seventeen States, mostly situated in the Southeast and Midwest, have death rates between 14 and 25 per million population. Twenty-eight States have fire death rates at or below the national fire death rate. While some State death rates are still high, overall, States have made great progress in lowering the absolute number of fire deaths and their deaths per million population.

Ten States, mostly large-population States, account for 51 percent of the national total U.S. fire deaths. Unless their fire problems are significantly reduced, the national total will be difficult to lower.

---

[1] The NFPA estimates fire deaths to be 3,430 in 2007. The 2007 NCHS mortality data were not available at the time *Fire in the United States* was under publication. Previous analyses of the NCHS mortality data suggest that fire deaths, on average, may be 8 to 9 percent higher.

[2] The fire death rate used throughout *Fire in the United States*, however, reflects the number of fire deaths (3,940) from the 2006 NCHS mortality data. This death rate is 13.2 fire deaths per million population.

# RESIDENCES AND OTHER PROPERTIES

Over the years, there has been little change in the proportion of fires, deaths, injuries, and dollar loss by the type of property involved. In terms of numbers of fires, the largest category continues to be outside fires (44 percent)—in fields, vacant lots, trash, etc. Many of these fires are intentionally set, but do not cause much damage. Residential and nonresidential structure fires together comprise 35 percent of fires, with residential structure fires outnumbering nonresidential structure fires by over three to one. What may surprise some is the large proportion of vehicle fires. In fact, nearly one out of every seven fires to which fire departments respond involves a vehicle.

By far, the largest percentage of deaths, 76 percent in 2007, occurs on residential properties, with the majority of these in one- and two-family properties. Vehicles account for the second largest percentage of fire deaths at 17 percent. Great attention is given to large, multiple-death fires in public places such as hotels, nightclubs, and office buildings. But the major attention-getting fires that kill 10 or more people are few in number and constitute only a small portion of overall fire deaths. Firefighters generally are doing a good job in protecting public properties in this country. Furthermore, these properties are generally required by local codes to have built-in fire suppression systems. The area with the largest problem is where it is least suspected—in people's homes. Prevention efforts continue to focus on home fire safety.

Only 4 percent of the 2007 fire deaths occur in commercial and public properties. Outside and other miscellaneous fires, including wildfires, are also a small factor (4 percent combined) in fire deaths.

The picture generally is similar for fire injuries, with 76 percent of all injuries occurring on residential properties. The remaining fire injuries are distributed across the other property types—nonresidential properties, 9 percent; vehicles, 7 percent; and outside and other fires, 9 percent.[3]

The picture changes somewhat for dollar loss. While residential properties are the leading property for dollar loss, nonresidential properties play a considerable role. These two property types account for 82 percent of all dollar loss. The proportion of dollar loss from outside fires may be understated because the destruction of trees, grass, etc., often is given zero value in fire reports if it is not commercial cropland or timber.

# CAUSES OF FIRES AND FIRE LOSSES
## *Residential*

At 40 percent, cooking is the leading cause of residential structure fires. Heating causes another 14 percent. These percentages (and those that follow) are adjusted, which proportionally spreads the unknown causes over the other 15 causes.

The two leading causes of residential fatal fires are smoking, at 18 percent, and other unintentional or careless actions, at 14 percent. The leading cause of residential fires that result in injuries is cooking (26 percent), followed by other unintentional or careless actions (11 percent) and open flame (also 11 percent). Cooking is, by far, the leading cause of fires with property loss, at 20 percent.

---

[3] Percentages do not add to 100 percent due to rounding.

## *Nonresidential*

For nonresidential structure fires, cooking is the leading cause of fires (23 percent), followed by intentionally set fires (12 percent).

The two leading causes of fatal fires in nonresidential structures are intentionally set fires, at 27 percent, and unintentionally set fires, or carelessness (19 percent). Cooking is the leading cause of fires that caused injuries (13 percent). Intentionally set fires and electrical malfunctions are the two leading causes of fires causing dollar loss, both at 13 percent.

Notably, intentionally set fires are the second leading cause of nonresidential structure fires, and are the leading causes of both fatal fires and dollar loss in nonresidential structures.

## *Vehicle*

Unintentionally set fires are the leading cause of fires, fatal fires, fires causing injuries, and fires causing dollar loss in vehicles (32, 48, 51, and 30 percent, respectively). Failure of equipment or heat source is the second leading cause in all categories except fatal fires (fires—24 percent, fires causing injuries—19 percent, fires causing dollar loss—27 percent) where the second leading cause is cause under investigation (28 percent).

## *Outside*

Intentionally set and unintentionally set fires are the leading causes of fatal outside fires, both at 37 percent. Unintentionally set fires are also the leading cause of fires, fires causing injuries, and fires causing dollar loss in outside fires (38, 53, and 38 percent, respectively). Undetermined causes after investigation are the second leading cause of outside fires and outside fires causing dollar loss (31 percent and 24 percent, respectively). Intentionally set fires are the second leading cause of outside fires causing injuries (23 percent).

## *Other*

Just as with vehicle and outside fires, unintentionally set fires are the leading cause of other fires, fatal fires, fires causing injuries, and fires causing dollar loss (41, 40, 66, and 40 percent, respectively). Failure of equipment or heat source is the second leading cause of other fires (27 percent), other fires causing injuries (11 percent), and other fires causing dollar loss (31 percent). In 24 percent of fatal fires, the cause of the fire is under investigation at the time the fire incident report was submitted. The cause is undetermined after the investigation in an additional 24 percent of fatal fires.

## RACE, AGE, AND GENDER CHARACTERISTICS OF VICTIMS

Fire losses affect all groups and races, rich and poor, North and South, urban and rural. But the problem is higher for some groups than for others. African-Americans and American Indian males have much higher fire death rates than the national average. African-Americans comprise a large and disproportionate share of total fire deaths, accounting for 22 percent of fire deaths—nearly twice as high as their share of the overall population (13 percent).

Approximately 50 percent more men die in fires than women. The reasons for this disparity are not known for certain. Suppositions include the greater likelihood of men being intoxicated and the more dangerous occupations of men (most industrial fire fatalities are males). Female fire deaths in the 70 and older age group account for nearly one-third of female fire deaths (31 percent). Male fire deaths, by contrast, are higher in the late midlife years (40 to 59). It is also known that men incur more injuries trying to extinguish the fire and rescue people than do women. Males aged 15 to 54 tend to have a slightly higher proportion of injuries, while young and older females have more injuries than males. Notably, older adult females have twice the proportion of fire injuries than older males.

The bulk of fire-related injuries occurs in adults aged 20 to 54. This age group accounts for over half of the fire injuries in 2007.

People with limited physical and cognitive abilities, especially older adults, are at a higher risk of death from fire than other groups.[4] Older adults account for 32 percent of fire deaths and 12 percent of estimated fire injuries.

As baby boomers enter retirement age, the demographic profile of the United States is expected to change dramatically. Over the coming decades, the older population will increase and a corresponding increase in fire deaths and injuries among older adults is likely.

In the past, children age 4 and younger were also considered to be at a high risk of death from fire, however, data indicate that the trend appears to be changing. The relative risk of children age 4 and younger dying in a fire is slightly less than that of the general population. In the future, additional analysis is required to confirm that this is a true trend change.

# CONCLUSIONS

This report shows that, overall, the fire problem in the United States continues to improve. Five-year fire loss rates are down. It is likely that several factors continue to contribute to these trends:

- Smoke alarms, whose usage has become nearly universal;
- Sprinklers, which quickly combat incipient fires, especially in nonresidential and multifamily buildings;
- Fire codes, which have been strengthened;
- Construction techniques and materials, which have been targeted specifically to fire prevention;
- Public education at the community, county, State, and Federal levels; and
- Improved firefighter equipment and training.

Even considering these positive trends, the United States still has a major fire problem compared to other industrialized Nations. The study and implementation of international fire prevention programs that have proved effective in reducing the number of fires and deaths should be considered.

Other areas that continue to be of concern:

- The very old remain at high risk of death from fire;
- The focus for injury prevention should be on adults aged 20 to 54;
- Certain ethnic groups remain at an enormous risk for fire deaths;

---

[4] USFA defines older adults as age 65 and older.

- Intentional fires are still a large problem in the United States, especially to outside and nonresidential properties;

- Contiguous States often have similar fire profiles; and

- Data challenges still exist: many records submitted to NFIRS by participating fire departments provide either incomplete or no information in some of the fields. Additionally, in preparing this report, it is assumed that participating fire departments have reported 100 percent of their fire incidents; however, this is not always the case. The completeness of all the information in the NFIRS modules will contribute to the refinement and confidence level of future analyses.

With continued improvements to the NFIRS system, data collection will also continue to improve. If we better understand the relative importance of the factors that lessen the fire problem, resources can be better targeted to have the most impact.

# Chapter 1
# Introduction

In 1973, the President's Commission on Fire Prevention and Control published *America Burning*. This document was the first in depth discussion of this country's fire problem, the most severe of the industrialized Nations. The report prompted a national awareness about the depth of the fire problem and the need for prevention efforts. By 1987, when a second commission was assembled, much progress had been made toward addressing the United States' fire problem. Among other objectives, *America Burning Revisited* redefined the strategies needed to further reduce loss of life and property to fire. As a direct result of these efforts and others like them, the U.S. fire problem no longer ranks as the most severe of the industrialized nations. Nonetheless, the United States continues to experience fire death rates and property losses from fire 2 to 2 1/2 times those of most of its sister Nations.[1] Many Americans are not aware of this or of the nature of the fire problem.

This report is a statistical portrait of fire in the United States. It is intended for use by a wide audience, including the fire service, the media, researchers, industry, government agencies, and interested citizens. The report focuses on the national fire problem. The magnitude and trends of the fire problem, the causes of fires, where they occur, and who gets hurt are topics that are emphasized.

This document is the Fifteenth major edition of *Fire in the United States* published by the U.S. Fire Administration (USFA). The previous editions have included

- First edition published in 1978; included 1975 and 1976 fire data.

- Second edition published in 1982; included 1977 and 1978 fire data.

- Third through fifth editions produced as working papers, but not published.

- Sixth edition published in 1987; included 1983 fire data.

- Seventh edition published in 1991; included 1983 to 1987 fire data.

- Eighth edition published in 1991; included 1983 to 1990 fire data.

- Ninth edition published in 1997; included 1985 to 1994 fire data, and focused on the residential/nonresidential fire problem and on firefighter casualties.

---

[1] "World Fire Statistics," Geneva Association Information Newsletter, October 2008, http://www.genevaassociation.org/PDF/WFSC/GA2008-FIRE24.pdf. As reported, the United States has a fire death rate of 1.41 fire deaths per 100,000 population for 2003 to 2005; Austria has the lowest comparable European death rate at 0.57 per 100,000 population. Switzerland is lower still but only includes fire deaths in buildings and excludes firefighter deaths.

- Tenth edition published in 1998; included 1986 to 1995 fire data, and provided a State-by-State profile of fires and an examination of firefighter casualties.

- Eleventh edition published in 1999; included 1987 to 1996 fire data, and focused on the residential/nonresidential fire problem and on firefighter casualties.

- Twelfth edition published in 2001; included 1989 to 1998 fire data; the last edition to use the National Fire Incident Reporting System (NFIRS) 4.1 data system, it included analyses of all of the previous topics under one cover: residential and nonresidential fire problems, State-by-State profiles, and firefighter casualties.

- Thirteenth edition published in 2004; included 1992 to 2001 fire data; the first edition to include the new NFIRS 5.0 data in the analyses; included the residential and nonresidential fire problem and firefighter casualties.

- Fourteenth edition published in 2007; included 1995 to 2004 fire data; primary focus on 2004; for the first time, only native NFIRS 5.0 data were used for NFIRS-based analyses; addressed the overall national fire problem; detailed analyses of the residential and nonresidential fire problem, firefighter casualties, and other subsets of the national fire problem were not included.

This Fifteenth edition covers the 5-year period 2003 to 2007, with a primary focus on 2007. As in the Fourteenth edition, only native NFIRS 5.0 data are used for NFIRS-based analyses.[2] This report addresses the overall national fire problem only. Detailed analyses of the residential and nonresidential fire problem, firefighter casualties, and other subsets of the national fire problem are addressed as separate, stand-alone publications. *Fire-Related Firefighter Injuries in 2004* was published in February 2008, *Residential Structure and Building Fires* was published in October 2008, and *Nonresidential Building Fires* will be available for downloading.

## SOURCES

The report is based primarily on NFIRS data, but uses other sources as well. Summary numbers for fires, deaths, injuries, and dollar loss are from the National Fire Protection Association's (NFPA's) annual survey of fire departments.[3] It also uses mortality data from the National Center for Health Statistics (NCHS)[4], population data from the U.S. Census Bureau, inflation adjustments from the Bureau of Labor Statistics' Consumer Price Index, and State statistics from State Fire Marshals' offices or their equivalents. Because the NCHS mortality data are based on a census or enumeration of deaths based on death certificates rather than an estimate, it is used as the primary source for the analysis of deaths in Chapter 2.

---

[2] Previous editions of *Fire in the United States* have presented 10-year trends. As many of the trends are based on national estimates that use the proportion of native NFIRS 5.0 data to allocate estimated fires and fire losses, trends in this edition are limited to 2003 and beyond when the proportion of native NFIRS 5.0 data exceeded 80 percent of the submitted data.

[3] The NFPA summary numbers are used for the overall U.S. fire losses; fire losses from vehicle, outside, and other fires; and as the basis for estimates of residential and nonresidential building fires. The alternative approach for these summary numbers is to use the relative percentage of fires (or other loss measures) from NFIRS and scale up (multiply by) the NFPA estimate of total fires. The results would be somewhat different from those using the NFPA subtotals. These differences are discussed in Appendix A. Better estimates of fire loss measures from NFIRS will not be available until a more robust method of estimation is developed.

[4] The NCHS data provide additional detail not available from the NFPA survey: State of fire death occurrence, age, gender, and race.

The most current year available for the NCHS mortality data is 2006.[5] Please note that for consistency's sake, national trend data are based on the NFPA survey estimates, not from the NCHS mortality data. The text notes where these data sources for fire deaths yield substantially different results.

The USFA gratefully acknowledges the use of the data and information provided by these groups. Data sources are cited for each graph and table in this report.

## National Fire Incident Reporting System

An indirect outgrowth of *America Burning*, the NFIRS was established in 1975 as one of the first programs of the National Fire Prevention and Control Administration, which later became the USFA. The basic concept of NFIRS has not changed since the system's inception. All States and all fire departments within them have been invited to participate on a voluntary basis. Participating fire departments collect a common core of information on an incident and any casualties that ensue by using a common set of definitions. The data may be written by hand on paper forms or entered directly through a computer. Local agencies forward the completed NFIRS modules to the State agency responsible for NFIRS data. The State agency combines the information with data from other fire departments into a statewide database and then transmits the data to the National Fire Data Center (NFDC) at USFA. Data on individual incidents and casualties are preserved incident by incident at local, State, and national levels. Once limited to fire incidents only, NFIRS now encompasses all incidents to which the fire department responds—fire, emergency medical services (EMS), hazardous materials (hazmat), and the like.

From an initial six States in 1976, NFIRS has grown in both participation and use. Over the life of the system, all 50 States, the District of Columbia, and more than 40 major metropolitan areas have reported to NFIRS. As well, more than 30,000 fire departments have been assigned participating NFIRS fire department identification numbers by their States. Approximately 1.3 million fire incident records and more than 18 million nonfire incident records are added to the database each year. NFIRS is the world's largest collection of incidents to which fire departments respond.

Between 1985 and 1999, the level of participation remained relatively constant: A few States came in or left the system each year, and at least 39 States reported to NFIRS. Most years also included participation from the District of Columbia. The number of fire departments participating within the States remained relatively constant as well, with a slight dip in participation during the system migration from version 4.1 to 5.0 in 1999. In 2000, the number of States increased to 43 and fire department participation began to bounce back from the version 5.0 transition low. Since 2000, State and fire department participation has been steadily increasing. In 2003, NFIRS reached the milestone of participation by all 50 States. The following year, NFIRS achieved another significant goal—NFIRS not only achieved the national goal of 100-percent State participation, including the District of Columbia, but for the first time, the Native American Tribal Authorities submitted data.

NFIRS continues to grow and mature. As of 2007, a new level of participation has been achieved: all 50 States, the District of Columbia, Native American Tribal Authorities, Northern Mariana Islands, and Puerto Rico all participated in NFIRS—for a total of 54 State, District, Tribal Authority, and Commonwealth entities (Table 2). Fire departments reporting fire incidents has grown to nearly 20,600[6] in 2007 (Figure 1).

---

[5] The 2007 NCHS data were not available at the time the analyses for *Fire in the United States* were undertaken; the 2006 NCHS mortality data were released in the late spring of 2009. As a result, the fire incident and fire injury analyses from NFIRS focus on 2007 while the fire death analyses are from 2006. As well, 5-year trends for the NCHS data are from 2002 to 2006 rather than the 2003 to 2007 trend data from NFPA and NFIRS.

[6] The 20,600 fire departments represent NFIRS fire incident submittal, regardless of the version of the data submitted (NFIRS 4.1 or NFIRS 5.0).

Across participating entities, an average of 59 percent of U.S. fire departments report fire incidents to NFIRS; 57 percent report in the new version (Table 3). With over half of all fire departments nationwide reporting fire incidents to NFIRS 5.0, the reporting departments represent a very large data set that enables USFA to make reasonable estimates of various facets of the fire problem. Participation in NFIRS is voluntary, although some States do require their departments to participate in the State system. Additionally, if a fire department is a recipient of a Fire Act Grant, participation is required.[7]

### Table 2. States Reporting Fire Incidents to NFIRS (1998-2007)

| State | 1998 | 1999 | 2000 | 2001 | 2002 | 2003 | 2004 | 2005 | 2006 | 2007 |
|---|---|---|---|---|---|---|---|---|---|---|
| Alabama | X | X | X | X | X | X | X | X | X | X |
| Alaska | X | X | X | X | X | X | X | X | X | X |
| Arizona | | | | X | X | X | X | X | X | X |
| Arkansas | X | X | X | X | X | X | X | X | X | X |
| California | X | | | | X | X | X | X | X | X |
| Colorado | | X | X | X | X | X | X | X | X | X |
| Connecticut | X | X | X | X | X | X | X | X | X | X |
| Delaware | | | | X | X | X | X | X | X | X |
| District of Columbia | X | X | | | | | X | X | | X |
| Florida | X | X | X | X | X | X | X | X | X | X |
| Georgia | X | X | X | X | X | X | X | X | X | X |
| Hawaii | X | X | X | X | X | X | X | X | X | X |
| Idaho | X | X | X | X | X | X | X | X | X | X |
| Illinois | X | X | X | X | X | X | X | X | X | X |
| Indiana | X | X | X | X | X | X | X | X | X | X |
| Iowa | X | X | X | X | X | X | X | X | X | X |
| Kansas | X | X | X | X | X | X | X | X | X | X |
| Kentucky | X | X | X | X | X | X | X | X | X | X |
| Louisiana | X | X | X | X | X | X | X | X | X | X |
| Maine | X | X | X | X | X | X | X | X | X | X |
| Maryland | X | X | X | X | X | X | X | X | X | X |
| Massachusetts | X | X | X | X | X | X | X | X | X | X |
| Michigan | X | X | X | X | X | X | X | X | X | X |
| Minnesota | X | X | X | X | X | X | X | X | X | X |
| Mississippi | | | X | X | X | X | X | X | X | X |
| Missouri | | X | X | X | X | X | X | X | X | X |
| Montana | X | X | X | X | X | X | X | X | X | X |
| Nebraska | X | X | X | X | X | X | X | X | X | X |
| Nevada | X | X | X | X | X | X | X | X | X | X |
| New Hampshire | X | | X | X | X | X | X | X | X | X |
| New Jersey | X | X | X | X | X | X | X | X | X | X |
| New Mexico | | | X | X | X | X | X | X | X | X |

---

[7] From the Assistance to Firefighters Grant Program guidance, if the applicant is a fire department, the department must agree to provide information, through established reporting channels, to NFIRS for the period covered by the assistance. If a fire department does not currently participate in the incident reporting system and does not have the capacity to report at the time of the award, the department must agree to provide information to the system for a 12-month period that begins as soon as the department develops the capacity to report. See http://www.firegrantsupport.com/docs/2009AFGguidance.pdf.

## Table 2. States Reporting Fire Incidents to NFIRS (1998-2007) – Continued

| State | 1998 | 1999 | 2000 | 2001 | 2002 | 2003 | 2004 | 2005 | 2006 | 2007 |
|---|---|---|---|---|---|---|---|---|---|---|
| New York | X | X | X | X | X | X | X | X | X | X |
| North Carolina | | | X | X | X | X | X | X | X | X |
| North Dakota | | | | X | X | X | X | X | X | X |
| Ohio | X | X | X | X | X | X | X | X | X | X |
| Oklahoma | X | X | X | X | X | X | X | X | X | X |
| Oregon | | X | X | X | X | X | X | X | X | X |
| Pennsylvania | | | | X | X | X | X | X | X | X |
| Rhode Island | | | | X | | X | X | | X | X |
| South Carolina | X | X | X | X | X | X | X | X | X | X |
| South Dakota | X | X | X | X | X | X | X | X | X | X |
| Tennessee | X | X | X | X | X | X | X | X | X | X |
| Texas | X | X | X | X | X | X | X | X | X | X |
| Utah | X | X | X | X | X | X | X | X | X | X |
| Vermont | X | X | X | X | X | X | X | X | X | X |
| Virginia | X | X | X | X | X | X | X | X | X | X |
| Washington | X | X | X | X | X | X | X | X | X | X |
| West Virginia | X | | | X | X | X | X | X | X | X |
| Wisconsin | X | X | X | X | X | X | X | X | X | X |
| Wyoming | X | X | X | X | X | X | X | X | X | X |
| Native American | | | | | | | X | X | X | X |
| Northern Mariana Islands | | | | | | | | | X | X |
| Puerto Rico | | | | | | | | X | X | X |
| Total | 40 | 40 | 43 | 49 | 49 | 50 | 52 | 52 | 53 | 54 |

Note:     Includes fire incidents submitted in both NFIRS versions 4.1 and 5.0.
Source:   NFIRS.

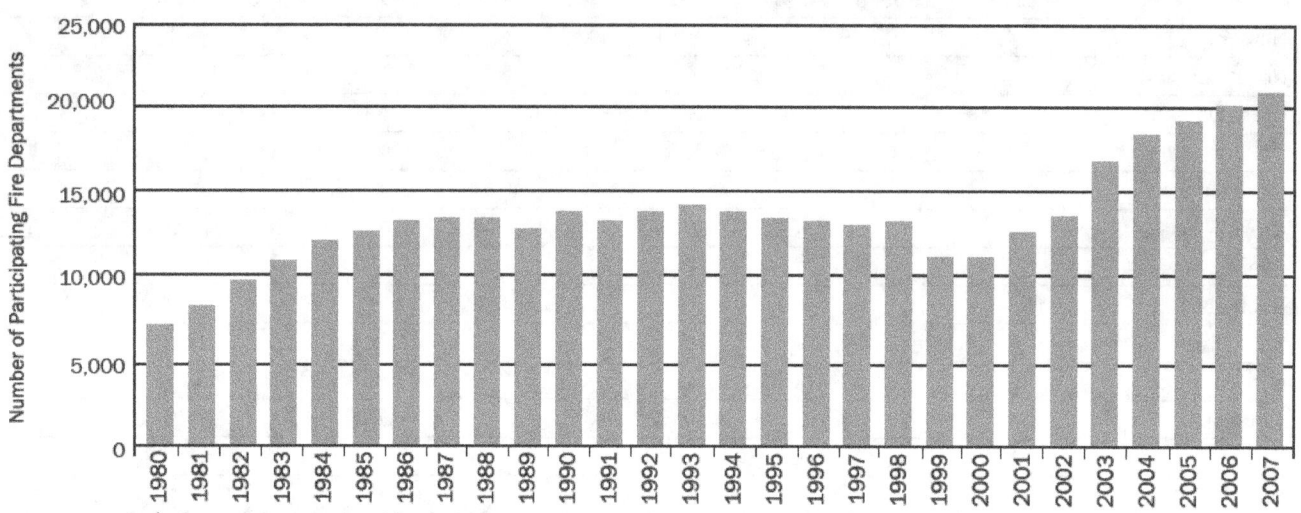

## Figure 1. NFIRS Fire Department Participation
### (1980-2007, fire incidents only)

Note:    Includes participation from NFIRS 4.1 and NFIRS 5.0.

Source:   NFIRS.

### Table 3. Fire Departments Reporting Fire Incidents to NFIRS in 2007

| State | No. of Fire Departments in State* | No. of Reporting Fire Departments (NFIRS 5.0) | Percent of Reporting Fire Departments (NFIRS 5.0) | No. of Reporting Fire Departments (All NFIRS) | Percent of Reporting Fire Departments (All NFIRS) |
|---|---|---|---|---|---|
| Alabama | 1,045 | 334 | 32% | 334 | 32% |
| Alaska | 210 | 101 | 48% | 101 | 48% |
| Arizona | 352 | 80 | 23% | 81 | 23% |
| Arkansas | 1,003 | 592 | 59% | 592 | 59% |
| California | 1,031 | 494 | 48% | 494 | 48% |
| Colorado | 405 | 238 | 59% | 239 | 59% |
| Connecticut | 261 | 231 | 89% | 231 | 89% |
| Delaware | 61 | 60 | 98% | 60 | 98% |
| District of Columbia | 1 | 1 | 100% | 1 | 100% |
| Florida | 671 | 378 | 56% | 399 | 59% |
| Georgia | 657 | 293 | 45% | 295 | 45% |
| Hawaii | 4 | 4 | 100% | 4 | 100% |
| Idaho | 247 | 168 | 68% | 168 | 68% |
| Illinois | 1,240 | 958 | 77% | 1000 | 81% |
| Indiana | 861 | 728 | 85% | 728 | 85% |
| Iowa | 875 | 324 | 37% | 324 | 37% |
| Kansas | 644 | 503 | 78% | 503 | 78% |
| Kentucky | 834 | 489 | 59% | 493 | 59% |
| Louisiana | 603 | 132 | 22% | 132 | 22% |
| Maine | 423 | 175 | 41% | 175 | 41% |

## Table 3. Fire Departments Reporting Fire Incidents to NFIRS in 2007–Continued

| State | No. of Fire Departments in State* | No. of Reporting Fire Departments (NFIRS 5.0) | Percent of Reporting Fire Departments (NFIRS 5.0) | No. of Reporting Fire Departments (All NFIRS) | Percent of Reporting Fire Departments (All NFIRS) |
|---|---|---|---|---|---|
| Maryland | 367 | 224 | 61% | 259 | 71% |
| Massachusetts | 365 | 337 | 92% | 337 | 92% |
| Michigan | 1,030 | 774 | 75% | 774 | 75% |
| Minnesota | 789 | 416 | 53% | 707 | 90% |
| Mississippi | 767 | 691 | 90% | 691 | 90% |
| Missouri | 896 | 549 | 61% | 549 | 61% |
| Montana | 432 | 184 | 43% | 184 | 43% |
| Nebraska | 483 | 221 | 46% | 221 | 46% |
| Nevada | 166 | 44 | 27% | 44 | 27% |
| New Hampshire | 243 | 157 | 65% | 159 | 65% |
| New Jersey | 745 | 608 | 82% | 609 | 82% |
| New Mexico | 366 | 184 | 50% | 184 | 50% |
| New York | 1,995 | 1,107 | 55% | 1,107 | 55% |
| North Carolina | 1,287 | 812 | 63% | 812 | 63% |
| North Dakota | 384 | 190 | 49% | 190 | 49% |
| Ohio | 1,200 | 1,143 | 95% | 1,174 | 98% |
| Oklahoma | 913 | 408 | 45% | 409 | 45% |
| Oregon | 327 | 240 | 73% | 240 | 73% |
| Pennsylvania | 2,389 | 739 | 31% | 739 | 31% |
| Rhode Island | 79 | 10 | 13% | 11 | 14% |
| South Carolina | 458 | 290 | 63% | 290 | 63% |
| South Dakota | 353 | 89 | 25% | 200 | 57% |
| Tennessee | 732 | 598 | 82% | 645 | 88% |
| Texas | 2,435 | 987 | 41% | 988 | 41% |
| Utah | 250 | 132 | 53% | 135 | 54% |
| Vermont | 237 | 165 | 70% | 165 | 70% |
| Virginia | 601 | 470 | 78% | 488 | 81% |
| Washington | 520 | 394 | 76% | 394 | 76% |
| West Virginia | 453 | 440 | 97% | 440 | 97% |
| Wisconsin | 860 | 430 | 50% | 430 | 50% |
| Wyoming | 132 | 84 | 64% | 84 | 64% |
| Native American | 100 | 7 | 7% | 7 | 7% |
| Northern Mariana Islands | 1 | 1 | 100% | 1 | 100% |
| Puerto Rico | 1 | 1 | 100% | 1 | 100% |
| **Total** | **33,784**** | **19,409** | **57%** | **20,022** | **59%** |

*The number of State fire departments was taken from the Fourteenth edition of *Fire in the United States* with the addition of the Northern Mariana Islands and Puerto Rico.

**This total differs from the 2007 NFPA estimate of 30,185 fire departments. The NFPA estimate is the official estimate used by USFA as its benchmark for the National Fire Department Census.

Sources: NFIRS (2007) and State Fire Marshals' offices or equivalent organizations (2006).

Corresponding to increased participation, the numbers of fires, deaths, and injuries, as well as estimates of dollar loss reported to NFIRS, also have grown—an estimated 63 percent of all U.S. fires to which fire departments responded in 2007 were captured in NFIRS.

There are, of course, many problems in assembling a real-world database, and NFIRS is no exception. Although NFIRS does not represent 100 percent of incidents reported to fire departments each year, the enormous data set and good efforts by the fire service result in a huge amount of useful information. Because of advances in computer technology and data collection techniques over the past 35 years and improvements suggested by participants, NFIRS has been revised periodically. The newest revision, NFIRS 5.0, became operational in January 1999.

NFIRS 5.0 captures information on all incidents, not just fires, to which a fire department responds. In addition to many data coding improvements, version 5.0 provides modules that recognize the increasingly diverse activities of fire departments today—an EMS Module, a Wildland Fire Module, an Apparatus Module, a Personnel Module, a Hazmat Module, and an Arson Module. The Basic Incident and Fire Modules of NFIRS 5.0 collect data in a different fashion than the precursor NFIRS systems.

The design of NFIRS 5.0 makes the system easier to use than previous NFIRS versions because it captures only the data required to profile the extent of the incident. Some fires, for example, require just basic information to be recorded, whereas others require considerably more detail.

State participation is voluntary, and each State specifies requirements for its fire departments. States have the flexibility to adapt their State reporting systems to their specific needs. As a result, the design of a State's data collection system varies from State to State. NFIRS 5.0 was designed so that data from State systems can be converted to a single format that is used at the national level to aggregate and store NFIRS data.

One of the most important changes is in the data format itself. All data in the system, regardless of its entry mechanism, are in NFIRS 5.0 format; non-NFIRS 5.0 data are converted to the 5.0 format. The proportion of native 5.0 data has steadily increased since the introduction of NFIRS 5.0 in 1999 (Table 4). At the time of publication, this proportion rose to 99 percent in the preliminary 2008 data. As of January 1, 2009, NFIRS 4.1 data are no longer accepted by the system. Prior to 2009, NFIRS 4.1 data in its converted form have been accepted by the system, however, USFA only uses native 5.0 data in its analyses.

### Table 4. NFIRS Fire Incident Data Reporting by Version (percent)

| Year | NFIRS 4.1 (converted to 5.0 format) | Native NFIRS 5.0 |
|------|-------------------------------------|------------------|
| 1999 | 92% | 8% |
| 2000 | 77% | 23% |
| 2001 | 48% | 52% |
| 2002 | 31% | 69% |
| 2003 | 19% | 81% |
| 2004 | 11% | 89% |
| 2005 | 5% | 95% |
| 2006 | 5% | 95% |
| 2007 | 2% | 98% |
| 2008* | 1% | 99% |

*Preliminary, based on the July 2009 monthly status data report.
Source: NFIRS.

## Uses of NFIRS

NFIRS data are used extensively at all levels of government for major fire protection decisions. At the local level, incident and casualty information is used for setting priorities and targeting resources. The data collected are particularly useful for designing fire prevention and educational programs and EMS-related activities specifically suited to the real emergency problems local communities face.

At the State level, NFIRS is used in many capacities. One valuable contribution is that some State legislatures use these data to justify budgets and to pass important bills on fire-related issues such as sprinklers, fireworks, and arson. Many Federal agencies, in addition to USFA, make use of NFIRS data. NFIRS data are used, for example, by the Consumer Products Safety Commission (CPSC) to identify problem products and to monitor corrective actions. The Department of Transportation (DOT) uses NFIRS data to identify fire problems in automobiles, which has resulted in mandated recalls. The Department of Housing and Urban Development (HUD) uses NFIRS to evaluate safety of manufactured housing (mobile homes). The USFA uses the data to design prevention programs, to order firefighter safety priorities, to assist in the development of training courses at the National Fire Academy (NFA), and for a host of other purposes. Thousands of fire departments, scores of States, and hundreds of industries have used the data. The potential for even greater use remains. One of the purposes of this report is to provide some idea of the types of information available from NFIRS. The information presented here is highly summarized; much more detail is available. The USFA report, *Uses of NFIRS: The Many Uses of the National Fire Incident Reporting System*, further describes the uses of the data and is available online at *http://www.usfa.dhs.gov/applications/publications*.

## U.S. Fire Departments

The number of fire departments in each State is taken from data, provided by each State's Fire Marshal's office or the equivalent agency, published in the previous edition of *Fire in the United States*. The USFA also maintains a database of fire departments. The USFA established the National Fire Department Census and its subsequent database in the fall of 2001 when the USFA launched a nationwide campaign for voluntary registration of fire departments. By July 2009, approximately 26,250 fire departments have registered with the census—about 87 percent of the estimated number of U.S. fire departments.

The census database is intended for use by the fire protection and prevention communities, allied professions, the general public, and the USFA. USFA uses the database to conduct special studies, guide program decisionmaking, and improve direct communication with individual fire departments. The database provides a current directory of registered fire departments and includes basic information such as addresses, department types, Web site addresses (if applicable), number of fire department personnel, and number of stations. The survey also collects information on specialized services that is released only in summary format. For more information about the National Fire Department Census or to download the list of registered fire departments, visit *http://www.usfa.dhs.gov/applications/census*.

# METHODOLOGY

Each edition of *Fire in the United States* refines and improves upon the last. In this edition, as in previous ones, an attempt has been made to keep the data presentation and analysis as straightforward as possible. It is also the desire of the USFA to make the report widely accessible to many different users, so it avoids unnecessarily complex methodology. Throughout this report, the term *fire casualties* refers to deaths and injuries; the term *fire losses* collectively includes fire casualties and dollar loss.

## Analytic Issues and Considerations

There are several longstanding issues regarding how to analyze NFIRS data when it is neither as complete nor as accurate as desired. Other analytic issues are the result of changes in definitions and data collection procedures from NFIRS 4.1 to NFIRS 5.0. The sections below discuss how the analyses in this report address these various issues.

## National Estimates

With the exception of the NFPA estimates for total fires and for vehicle, outside, and other fires, the numbers in this report are scaled-up national estimates or percentages, not just the raw totals from NFIRS. The national estimates are derived by computing a percentage of fires, deaths, injuries, or dollar loss in a particular NFIRS category and multiplying it by the corresponding total number from the NFPA annual survey. For example, the national estimate for the number of injuries by age group used in the calculation for the fire injury rate per million population (Figure 14, Rate of Fire Injuries by Age and Gender) was computed by taking the percentage of NFIRS fire injuries (with known age) and multiplying it by the estimated total number of fire injuries from the NFPA survey. This methodology is the accepted practice of national fire data analysts.

Ideally, one would like to have all of the data come from one consistent data source. Because the "residential population protected" is not reported to NFIRS by many fire departments and the reliability of that data element is suspect in many other cases, especially where a county or other jurisdiction is served by several fire departments that each report their population protected independently, this data element was not used. Instead, extrapolations of the NFIRS sample to national estimates are made using the NFPA survey for the gross totals of fires, deaths, injuries, and dollar loss.

One problem with this approach is that the proportions of fires and fire losses differ between the large NFIRS sample and the NFPA survey sample. Nonetheless, to be consistent with approaches being used by other fire data analysts, the NFPA estimates of fires, deaths, injuries, and dollar loss are used as a starting point. The details of the fire problem below this level are based on proportions from NFIRS. Because the proportions of fires and fire losses differ between NFIRS and the NFPA estimates, from time to time, this approach leads to minor inconsistencies. These inconsistencies will remain until all estimates can be derived from NFIRS data alone.

## Unknown Entries

On a fraction of the incident reports or casualty reports sent to NFIRS, the desired information for many data items either is not reported or is reported as "unknown." The total number of blank or unknown entries is often larger than some of the important subcategories. For example, 43 percent of fatal structure fires reported in 2007 do not have sufficient data recorded in NFIRS to determine fire cause. The lack of data, especially for these fatal fires, masks the true picture of the fire problem. Many prevention and public education programs use NFIRS data to target at-risk groups or to address critical problems, fire officials use the data in decisionmaking that affects the allocation of firefighting resources, and consumer groups and litigators use the data to assess product fire incidence. When the unknowns are large, the credibility of the data suffers. In some cases, even after the best attempts by fire investigators, the information is truly unknown. In other cases, the information reported as unknown in the initial NFIRS report is not updated after the fire investigation is completed. Fire departments need to be more aware of the effect of incomplete data reporting and need to update the initial NFIRS report if additional information is available after investigation.

In making national estimates, the unknowns should not be ignored. The approach taken in this report is to provide not only the "raw" percentages of each category, but also the "adjusted" percentages computed using only those incidents for which data were provided. This calculation, in effect, distributes the fires for which the data are unknown in the same proportion as the fires for which the data are known, which may or may not be approximately right.

To illustrate using the cause of residential structure fires: Cooking was determined as the fire cause for 32.9 percent of reported fires in 2007; another 18.4 percent of reported fires had cause unknown. Thus, the percent of fires that had their cause reported was 100−18.4 = 81.6 percent. With the unknown causes proportioned like the known causes, the adjusted percent of cooking fires can then be computed as $32.9/81.6 = 40.3$ percent.

As in past editions of Fire in the United States, both the reported data and the adjusted data (if unknowns are present) are plotted on the bar charts in this edition.

## Incomplete Loss Reporting

As troublesome as insufficient data for the various NFIRS data items can be, equally challenging is the apparent nonreporting of injuries and property loss associated with many fire incidents. For example, there are many reported fires where the flame spread indicates damage but property loss is not reported. It is notoriously difficult to estimate dollar loss, but an approximation is more useful than leaving the data item blank. The degree to which there is incomplete reporting of civilian fire deaths is more difficult to identify, as the numbers of deaths are relatively small. Incomplete reporting of civilian injuries also is difficult to ascertain, but the injury-per-fire profiles for most departments are within reason.

## Computing Trends

A frequently asked question is how much a particular aspect of the fire problem has changed over time. The usual response is in terms of a percentage change from one year to another. As we are dealing with real-world data that fluctuate from year to year, a percent change from one specific year to another can be misleading. This is especially true when the beginning and ending data points are extremes—either high or low. For example, Table 5 shows that the percent change from 380 fire injuries in multifamily buildings in 2003 to 405 fire injuries in 2007 would be an increase of 6.6 percent. Yet, if we were to choose 2005 as the beginning data point (510 fire injuries), this change would show a substantial 20.6 percent *decrease*. As we are interested in *trends* in the U.S. fire problem, Fire in the United States presents the computed best-fit linear trend line (which smoothes fluctuations in the year-to-year data) and presents the change over time based on this trend line. The overall 5-year trend is an increase in injuries of 6.7 percent which, in this example, closely matches the point-to-point change between the 5 years. As noted above, trends that incorporate NFIRS data from the 5.0 system may have subtle changes as a result of the system design and not a true trend change.

### Table 5. Comparison of Percentage Change Indicators

| Year | Multifamily Building Fire Injuries | Best-Fit Linear Trend | Change between 2003 and 2007 | Change between 2005 and 2007 |
|---|---|---|---|---|
| 2003 | 380 | 419 | 380 | |
| 2004 | 425 | 426 | | |
| 2005 | 510 | 433 | | 510 |
| 2006 | 445 | 440 | | |
| 2007 | 405 | 447 | 405 | 405 |
| Percent change | | 6.7% | 6.6% | -20.6% |

Source: NFPA.

## Rounding

Percentages on each chart are rounded to one decimal point. Textual discussions cite these percentages as whole numbers. Thus, 13.4 percent is rounded to 13 percent and 13.5 percent is rounded to 14 percent. National estimates are rounded as follows: fires are rounded to the nearest 100 fires, deaths to the nearest 5 deaths, injuries to the nearest 25 injuries, and loss to the nearest million dollars.

## *Representativeness of the Sample*

The percentage of fire departments participating in NFIRS varies from State to State, with some States not participating at all in some years. To the best that USFA can determine, the distribution of participants is reasonably representative of the entire Nation, even though the sample is not random. The data set is so large—63 percent of all fires—and reasonably distributed geographically and by size of community that there is no known major bias that will affect the results.

In a joint study effort, USFA and NFPA examined the biases in NFIRS participation, specifically whether the fire experience of NFIRS-reporting departments differed systematically from the fire experience of other nonreporting departments within the same population. Results based on data from 1997 and 2002 indicated that there were differences in total fire loss estimates derived from NFIRS reporting departments and non-NFIRS reporting departments; however, the degree of difference was not great enough to merit adjusting current scaling methodologies. Thus, this edition of *Fire in the United States* will continue to use the long-standing methodology of scaling NFIRS estimates with NFPA total fire estimates.

In the fall of 2008, as required by the U.S. Office of Management and Budget, USFA undertook a study of the NFIRS data set to examine the potential bias in NFIRS due to fire department nonresponse. As a result, USFA completed an analysis to identify fire departments that do not participate in NFIRS, characteristics of these departments, and whether their nonresponse impacts the representativeness of NFIRS. Undertaken on a regional and county basis, the analysis provided insight into what, if any, adjustments could be made to minimize the impact of possible reporting bias on the fire loss estimates. States of particular concern for nonreporting are located in the Northeast and West regions of the country where the average rates of reporting are approximately 72 percent for each of these regions. By contrast, the Midwest region has an estimated 87 percent reporting rate.

Most of the NFIRS data exhibit stability from 1 year to another, without radical changes. Results based on the full data set are generally similar to those based on part of the data, another indication of data reliability. Although improvements could be made—the individual incident reports could and should be filled out more completely and more accurately than they are today (as can be said about most real-world data collections as large as NFIRS), and all participating departments should have the same reporting requirements—the overall portrayal is a reasonably accurate description of the fire situation in the United States.

## *Comparing Statistics to Past Editions*

Differences between the current NFIRS and older versions have, or may have, an effect on the analyses of fire topics. These differences, the result of both coding changes and data element design changes, required revisions to long-standing groupings and analyses. The revisions have caused some challenges when comparing current data to past data.

## Data Collection and Reporting Changes

Streamlined reporting for qualified incidents; the collection of smoke alarm and Automatic Extinguishing System (AES) data (formerly called sprinklers); definition changes for some property types; the differentiation between buildings and structures; and changes in the cause methodology are among the areas that are approached differently in NFIRS 5.0.[8] As these revisions have resulted in changes in overall trends—some subtle and some substantial—this edition does not include trends based on previous versions of NFIRS data. Subsequent editions will build on the analyses presented here.

## Confined Fires

The limited reporting of confined, low-loss structure fires allows the fire service to capture incidents that either might have gone unreported prior to the introduction of NFIRS 5.0 or were reported, but as a nonfire incident, as no loss was involved.[9] Data from this reporting option for structure fires were investigated in a 2006 USFA report, *Confined Structure Fires*. The addition of these fires results in increased proportions of cooking and heating fires in analyses of structure fire cause. In other analyses, the inclusion of confined fires may result in larger unknowns than in previous editions of this report, as detailed reporting of fire specifics is not required. In 2007, confined fires accounted for 19 percent of all fires and 44 percent of structure fires. Seventy-nine percent of confined structure fires were no- or low-loss cooking fires (61 percent) and heating fires (18 percent).

## Definitional Changes

### Property Types

Examples of property type changes include manufacturing and properties that are vacant and under construction. Manufacturing properties are no longer assigned a specific property use code based on the type of item manufactured. Instead, these properties are differentiated by an additional data element, "on-site materials." Vacant and under construction now is an attribute of a structure and no longer is considered a separate property type.

### Buildings and Structures

NFIRS 5.0 allows for the differentiation between buildings and nonbuildings. In NFIRS 5.0, a structure is a built object and can include platforms, tents, connective structures (e.g., bridges), and various other structures (e.g., fences, underground work areas, etc.). This distinction between buildings and nonbuildings is important when determining the effectiveness of engineered fire safety features such as smoke alarms and AES. These important components of early fire detection and automatic suppression apply to buildings and not necessarily to other types of structures. To facilitate analysis of these components and to acknowledge that prevention efforts generally are focused on buildings, USFA separates the subset of buildings from the rest of the structures.

---

[8] Other changes between NFIRS 4.1 and 5.0, such as mutual aid, do not have as significant an impact on analyses. As such, they are not addressed here. The NFIRS 5.0 documentation at *http://www.nfirs.fema.gov/documentation* provides detailed information.

[9] Some States routinely reported such nonloss fires as smoke scares. The result, from a reporting viewpoint, is that the incident is reported, but not coded as a fire incident.

Structure fires are defined by the NFIRS incident type. Structure fires are defined as the 110 incident type series (structure fires) and the 120 incident type series (fires in mobile property used as a fixed structure).[10] These incident types are:

- o   111 - Building fire;
- o   112 - Fires in structure other than in a building;[11]
- o   113 - Cooking fire, confined to container;
- o   114 - Chimney or flue fire, confined to chimney or flue;
- o   115 - Incinerator overload or malfunction, fire confined;
- o   116 - Fuel burner/boiler malfunction, fire confined;
- o   117 - Commercial compactor fire, confined to rubbish;
- o   118 - Trash or rubbish fire, contained;
- o   120 - Fire in mobile property used as a fixed structure, other;
- o   121 - Fire in mobile home used as fixed residence;
- o   122 - Fire in motor home, camper, recreational vehicle; and
- o   123 - Fire in portable building, fixed location.

Building fires are a subset of structure fires. They are defined as structure fires where the structure type is an enclosed building, a fixed portable, or mobile structure. By definition, this excludes nonbuilding structures. Previous USFA analyses demonstrated that confined structure fire incidents with full incident reporting primarily occurred in buildings. To accommodate the confined fire incident types with abbreviated incident reporting, the incident is also assumed to be a building if the structure type is not specified. In terms of NFIRS data, building fires are therefore defined as:

- •   NFIRS version 5.0 data.

- •   Aid types:
    - o   1 - Mutual aid received;
    - o   2 - Automatic aid received; and
    - o   5 - Other aid given.

    Note:  Mutual aid given and automatic aid given (aid types 3 and 4) were excluded to avoid double counting of incidents.

- •   Incident types:
    - o   111 - Building fire;
    - o   112 - Fires in structure other than in a building;[12]
    - o   113 - Cooking fire, confined to container;
    - o   114 - Chimney or flue fire, confined to chimney or flue;
    - o   115 - Incinerator overload or malfunction, fire confined;
    - o   116 - Fuel burner/boiler malfunction, fire confined;
    - o   117 - Commercial compactor fire, confined to rubbish;
    - o   118 - Trash or rubbish fire, contained;
    - o   120 - Fire in mobile property used as a fixed structure, other;
    - o   121 - Fire in mobile home used as fixed residence;

---

[10] Note that incident type 110 is not included. Incident type 110 is a conversion code for NFIRS 4.1. Incident type 110 is not a valid code for data collected in NFIRS 5.0. Incidents in the NFIRS 5.0 database with a 110 incident type are incidents collected under the NFIRS 4.1 system and are converted to NFIRS 5.0 compatible data.

[11] Preliminary findings noted that the fires coded as 112 appear to be buildings. A more detailed look at these incident types is required to determine whether they were coded correctly.

[12] See footnote 11.

    o  122 - Fire in motor home, camper, recreational vehicle; and

    o  123 - Fire in portable building, fixed location.

Note: Incident types 113 to 118 do not specify if the structure is a building.

- Structure type:
  - o  1 - Enclosed building;
  - o  2 - Fixed portable or mobile structure; and
  - o  structure type not specified (null entry).

## Cause Methodology

Since the introduction of NFIRS version 5.0, the implementation of the cause hierarchy has resulted in a steady increase in the percentages of unknown fire causes. This increase may be due, in part, to the fact that the original cause hierarchy (described in *Fire in the United States 1995-2004, Fourteenth Edition*) does not apply as well to version 5.0. Causal information now collected as part of NFIRS version 5.0 was not incorporated in the old hierarchy. As a result, many incidents were assigned to the unknown cause category. As the hierarchy was originally designed for structures, incidents that did not fit well into the structure cause categories were also assigned to the unknown category.

### Structure Fires

To capture the wealth of data available in NFIRS 5.0, USFA developed a modified version of the previous cause hierarchy for structure fires as shown in Table 6. The revised schema provides three levels of cause descriptions: a set of more detailed causes (priority cause description), a set of mid-level causes (cause description), and a set of high-level causes (general cause description). The priority cause description and the cause description existed previously as part of the original cause hierarchy, but have been expanded to capture the new 5.0 data.

The causes of fires are often a complex chain of events. To make it easier to grasp the "big picture," the 16 mid-level categories of fire causes such as heating, cooking, and playing with heat source are used by the USFA here and in many other reports. The alternative is to present scores of detailed cause categories or scenarios, each of which would have a relatively small percentage of fires. For example, heating includes subcategories such as misuse of portable space heaters, wood stove chimney fires, and fires involving gas central heating systems. Experience has shown that the larger categories are useful for an initial presentation of the fire problem. A more detailed analysis can follow.

Fires are assigned to one of the 16 mid-level cause groupings using a hierarchy of definitions, approximately as shown in Table 7.[13] A fire is included in the highest category into which it fits on the list. If it does not fit the top category, then the second one is considered, and if not that one, the third, and so on. (See Table 6 Note for examples.)

### Vehicle, Outside, and Other Fires

While these new cause categories have usefulness for the other property types—vehicle, outside, and other fires—there are limitations. USFA plans to investigate and develop specific cause categories for vehicle, outside, and other fires. Until then, the causes of fires for these property types presented in this edition are based on the distributions for the cause of ignition data element. This data element captures a very broad sense of the cause of the fire.

---

[13] The structure fire cause hierarchy and specific definitions in terms of the NFIRS 5.0 codes may be found at http://www.usfa. dhs.gov/fireservice/nfirs/tools/fire_cause_category_matrix.shtm. The hierarchy involves a large number of subcategories that are later grouped into the 16 mid-level cause categories, then the 8 high-level cause groupings.

## Table 6. Three-Level Structure Fire Cause Hierarchy

| Priority Cause Description (in hierarchical order) | Cause Description | General Cause Description |
|---|---|---|
| Exposure | Exposure | Exposure |
| Intentional | Intentional | Firesetting |
| Investigation with Arson Module | Investigation with Arson Module | Unknown |
| Children Playing | Playing with Heat Source | Firesetting |
| Other Playing | | |
| Natural | Natural | Natural |
| Fireworks | Other Heat | Flame, Heat |
| Explosives | | |
| Smoking | Smoking | |
| Heating | Heating | Equipment |
| Cooking | Cooking | |
| Air Conditioning | Appliances | |
| Electrical Distribution | Electrical Malfunction | Electrical |
| Appliances | Appliances | Equipment |
| Special Equipment | Other Equipment | |
| Processing Equipment | | |
| Torches | Open Flame | Flame, Heat |
| Service Equipment | Other Equipment | Equipment |
| Vehicle, Engine | | |
| Unclassified Fuel-Powered Equipment | | |
| Unclassified Equipment w/ Other or Unknown Fuel Source | Unknown | Unknown |
| Unclassified Electrical Malfunction | Electrical Malfunction | Electrical |
| Matches, Candles | Open Flame | Flame, Heat |
| Open Fire | | |
| Other Open Flame, Spark | Other Heat | |
| Friction, Hot Material | | |
| Ember, Rekindle | Open Flame | |
| Other Hot Object | Other Heat | |
| Natural Condition, Other | Natural | Natural |
| Heat Source or Product Misuse | Other Unintentional, Careless | Unknown |
| Equipment Operation Deficiency | Equipment Misoperation, Failure | Equipment |
| Equipment Failure, Malfunction | | |
| Trash, Rubbish | Unknown | Unknown |
| Other Unintentional | Other Unintentional, Careless | |
| Exposure (Fire Spread, Other) | Exposure | Exposure |
| Unknown | Unknown | Unknown |

Note:    Fires are assigned to a cause category in the hierarchical order shown. For example, if the fire is judged to be intentionally set and a match was used to ignite it, it is classified as intentional and not open flame because intentional is higher on the list.

## Deaths, Injuries, and Dollar Loss

In past editions of *Fire in the United States*, the cause sections have included the distributions of deaths, injuries, and dollar loss by fire cause. In principle, it is the cause of the **fire** which results in deaths, injuries, and dollar loss that should be analyzed, not numbers of deaths and injuries associated with fire causes. Therefore, in this edition, analyses of fire cause will address fires that cause deaths (fatal fires), fires that cause injuries, and fires that cause dollar loss.

## Other Considerations

An additional problem to keep in mind when considering the rank order of causes in this report is that sufficient data to categorize the cause were not reported to NFIRS for all fatal fires in the database. The rank order of causes might be different than shown here if the cause profile for the fires where causes were not reported to NFIRS were substantially different from the profile for the fires where causes were reported. However, there is no information available to indicate that there is a major difference between the known causes and the unknown causes, and so our present best estimate of fire causes is based on the distribution of the fires with known causes.

NFIRS fire causal data can be analyzed in many ways, such as by the heat source, equipment involved in ignition, factors contributing to ignition, or many other groupings. The hierarchy of causes used in this report has proven to be useful in understanding the fire problem and targeting prevention, but other approaches are useful too. Because the NFIRS database stores records fire-by-fire, and not just in summary statistics, a wide variety of analyses is possible.

The cause categories displayed in the graphs are listed in the same order to make comparisons easier from one to another. The y-scale varies from figure to figure depending on the largest percentage that is shown; the y-scale on a figure with multiple charts, however, is always the same.

## Table 7. Mid-Level Cause Grouping

| Cause Category | Definition |
|---|---|
| Exposure | Caused by heat spreading from another hostile fire |
| Intentional | Cause of ignition is intentional or fire is deliberately set |
| Investigation with Arson Module | Cause is under investigation and a valid NFIRS arson module is present |
| Playing with Heat Source | Includes all fires caused by individuals playing with any materials contained in the categories below as well as fires where the factors contributing to ignition include playing with heat source. Children playing with fire are included in this category |
| Natural | Caused by the sun's heat, spontaneous ignition, chemicals, lightning, static discharge, high winds, storms, high water including floods, earthquakes, volcanic action, and animals |
| Other Heat | Includes fireworks, explosives, flame/torch used for lighting, heat or spark from friction, molten material, hot material, heat from hot or smoldering objects |
| Smoking | Cigarettes, cigars, pipes, and heat from undetermined smoking materials |
| Heating | Includes confined chimney or flue fire, fire confined to fuel burner/boiler malfunction, central heating, fixed and portable local heating units, fireplaces and chimneys, furnaces, boilers, water heaters as source of heat |
| Cooking | Includes confined cooking fires, stoves, ovens, fixed and portable warming units, deep fat fryers, open grills as source of heat |
| Appliances | Includes televisions, radios, video equipment, phonographs, dryers, washing machines, dishwashers, garbage disposals, vacuum cleaners, hand tools, electric blankets, irons, hairdryers, electric razors, can openers, dehumidifiers, heat pumps, water cooling devices, air conditioners, freezers and refrigeration equipment as source of heat |
| Electrical Malfunction | Includes electrical distribution, wiring, transformers, meter boxes, power switching gear, outlets, cords, plugs, surge protectors, electric fences, lighting fixtures, electrical arcing as source of heat |
| Other Equipment | Includes special equipment (radar, x-ray, computer, telephone, transmitters, vending machine, office machine, pumps, printing press, gardening tools, or agricultural equipment), processing equipment (furnace, kiln, other industrial machines), service, maintenance equipment (incinerator, elevator), separate motor or generator, vehicle in a structure, unspecified equipment |
| Open Flame, Spark (Heat From) | Includes torches, candles, matches, lighters, open fire, ember, ash, rekindled fire, backfire from internal combustion engine as source of heat |
| Other Unintentional, Careless | Includes misuse of material or product, abandoned or discarded materials or products, heat source too close to combustibles, other unintentional (mechanical failure/malfunction, backfire) |
| Equipment Misoperation, Failure | Includes equipment operation deficiency, equipment malfunction |
| Unknown | Cause of fire undetermined or not reported |

Source: USFA.

## *Differences Between NFIRS Data and NFPA Survey Data*

As there are differences between any two analysts using NFIRS data because of the many assumptions and decisions about how to analyze incomplete and imperfect data, there can be inconsistencies between different data sources. In particular, there are discrepancies between the NFIRS 5.0 data and the NFPA annual survey data, especially with the 2007 data. With the exception of the NFPA survey data for 2007, NFIRS 5.0 and NFPA both show declines in deaths and injuries per 1,000 fires, but at different rates. Again with the exception of 2007, NFIRS 5.0 dollar loss per fire is lower than that of NFPA.[14] This issue is discussed further in Appendix A.

## *Unreported Fires*

NFIRS only includes fires to which the fire service responded. In some States, fires attended by State fire agencies (such as forestry) are included; in other States, they are not.

### Nonreporting to NFIRS

NFIRS includes fires from all States, but does not include incidents from many fire departments within participating States—the percent of fire departments reporting varies greatly from State to State. However, if the fires from the reporting departments are reasonably representative, this omission does not cause a problem in making useful national estimates for any but the smallest subcategories of data and for some geographic analyses.

Some fire departments submit information on most, but not all, of their fires. Sometimes the confusion is systematic, as when no-loss cooking fires or chimney fires are not reported. Sometimes it is inadvertent, such as when incident reports are lost or accidentally not submitted. The information that is received is assumed to be the total for the department and is extrapolated as such. Although there was no measure of the extent of this problem in the past, the NFIRS 5.0 provides fire departments with the capability to report this information in a simplified, more straightforward manner.

### Nonreporting to the Fire Service

A very large number of fires are not reported to the fire service at all. Most are believed to be small fires in the home or in industry that go out by themselves or are extinguished by the occupant. Special surveys of homes and businesses are needed to estimate the unreported fires. No attempt is made here to estimate them. Studies undertaken in the mid-1970s and again in the mid-1980s on unreported residential fires indicated that a substantial number of fires are not reported to local fire departments. The 1984 Consumer Product Safety Commission (CPSC) study on unreported residential fires noted that, of the estimated number of fires in residences, only 3 percent were reported to fire departments and 97 percent were not.[15] Although the vast majority of fire incidents are unreported because they are small, confined, and immediately extinguished, they are still fires. Even the largest fire starts small; hence, all fires regardless of size, merit prevention attention, and analytic investigation.

---

[14] As NFIRS 5.0 now captures a large number of small, low-loss fires (confined fires) thought to be unreported previously, these differences in loss rates per fire may not be surprising.

[15] *1984 National Sample Survey of Unreported, Residential Fires, Final Technical Report* prepared for the U.S. Consumer Product Safety Commission, Contract No. C-83-1239, Audits & Surveys, Inc., Princeton, NJ (1985).

## ORGANIZATION OF REPORT

Chapter 2 presents an overview of the national fire problem in terms of the total numbers of fires, deaths, injuries, and dollar loss (the four principal measures used to describe the fire problem). In a slight departure from the last edition of *Fire in the United States*, Chapter 3 provides an overview of buildings (residential and nonresidential) rather than structures in general. This change focuses the analysis on the 93 percent of structure fires that occur in buildings where individuals live and work. Vehicle and other mobile properties and outside and other properties complete the chapter.

Appendix A discusses the differences between NFPA and NFIRS data.

Appendix B presents 10-year trends for the national fire problem.

Appendix C presents 10-year trends for the major property use categories: residential structures, nonresidential structures, vehicle, outside, and other.

Most of the data are presented graphically for ease of comprehension. The specific data associated with the graphs are provided directly with the chart.

This edition of *Fire in the United States* concludes with a topical index.

The residential and nonresidential fire problems, respectively, have been published as stand-alone reports with more detail. Firefighter casualties are published in two reports: the annual *Firefighter Fatalities in the United States* report and a companion report, *Fire-Related Firefighter Injuries*, last updated with 2004 data. These reports and other related resources can be found at the following URL: *http://www.usfa.dhs.gov/ statistics/reports/fius.shtm* under the section: "Other Resources on the Fire Problem."

# Chapter 2
# The National Fire Problem

The United States has a severe fire problem, more so than is generally perceived. Nationally, there are millions of fires, thousands of deaths, tens of thousands of injuries, and billions of dollars lost—which make the U.S. fire problem one of great national importance. The indirect costs of fire increase the magnitude of economic loss tenfold.

The United States had a yearly average of 1,587,000 fires and 3,635 fire deaths from 2003 to 2007 (Figure 2.)[1,2] In terms of estimates of fires, fire deaths, and fire injuries, the estimates are lower than they were 5 years ago. The United States has made much progress in the intervening years since the USFA was established, but continues to have fire death rates and property losses that are among the largest of the industrialized Nations.

Fire injury statistics in Figure 2 are not as clear-cut as the fire death totals because of ambiguity about the completeness of defining and reporting minor injuries. In addition, many injured people go directly to a medical care facility themselves without a report to, or being treated by, the fire department or local emergency medical services (EMS) responders. Civilian injuries from fires averaged 17,600 per year over the 5-year period.[3] Firefighter injuries averaged 39,885 from those fires.[4] Past studies suggest that the number of civilian injuries associated with fires that are not reported to the fire service might be several times that of the number from reported fires, as discussed in Chapter 1. Fire-caused injuries to civilians declined by 5 percent over the 5 years, despite a 4-percent increase in the national population over this period.[5]

---

[1] Fires are rounded to the nearest 1,000; deaths to the nearest 5.

[2] The NCHS mortality data and the NFPA fire death estimates yield substantially different averages. Based on the NFPA survey data, fire deaths averaged 3,635 between 2003 and 2007. Because the NCHS mortality data lags, the latest 5-year average, 2002 to 2006 is 3,986.

[3] Injuries are rounded to the nearest 100.

[4] National Fire Protection Association, Firefighter Injuries by Type of Duty, http://www.nfpa.org/itemDetail.asp?categoryID=955 &itemID=23466&URL=Research/Fire%20statistics/The%20U.S.%20fire%20service&cookie%5Ftest=1. The 5-year average of firefighter fireground injuries is derived from summary data presented in this table.

[5] U.S. Census Bureau, Population Estimates, http://www.census.gov/popest/states/tables/NST-EST2008-01.xls.

## Figure 2. Fires and Fire Losses (2003-2007)

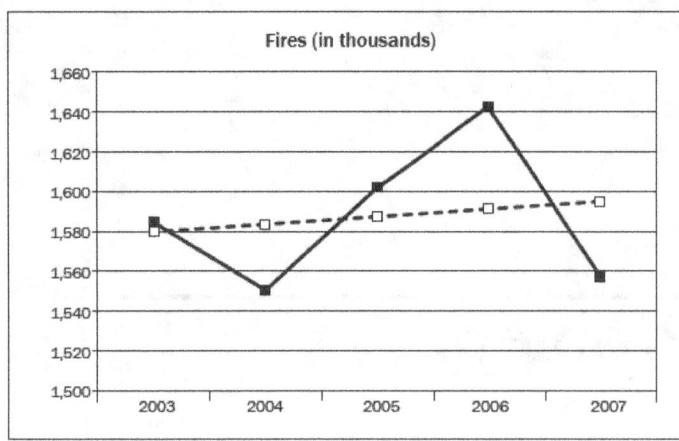

| FIRES (THOUSANDS) | |
| --- | --- |
| Year | Value |
| 2003 | 1,584.5 |
| 2004 | 1,550.5 |
| 2005 | 1,602.0 |
| 2006 | 1,642.5 |
| 2007 | 1,557.5 |
| 5-Year Trend (%) | 1.0% |

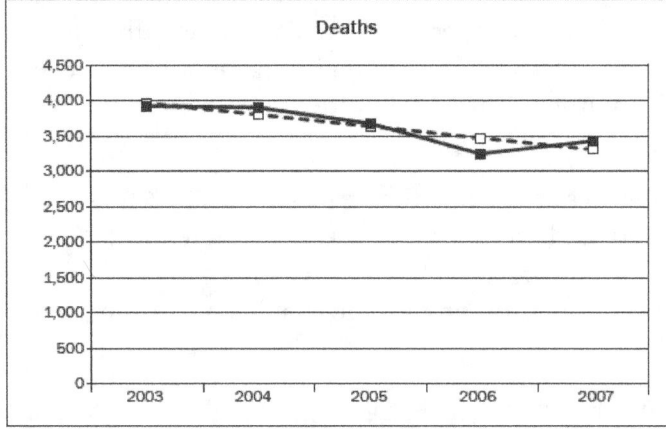

| DEATHS | |
| --- | --- |
| Year | Value |
| 2003 | 3,925 |
| 2004 | 3,900 |
| 2005 | 3,675 |
| 2006 | 3,245 |
| 2007 | 3,430 |
| 5-Year Trend (%) | -16.6% |

Note: Fire Deaths from NCHS are higher than the estimates from NFPA shown here. The NCHS 5-year (2002-2006) trend shows a decrease of 0.4 percent

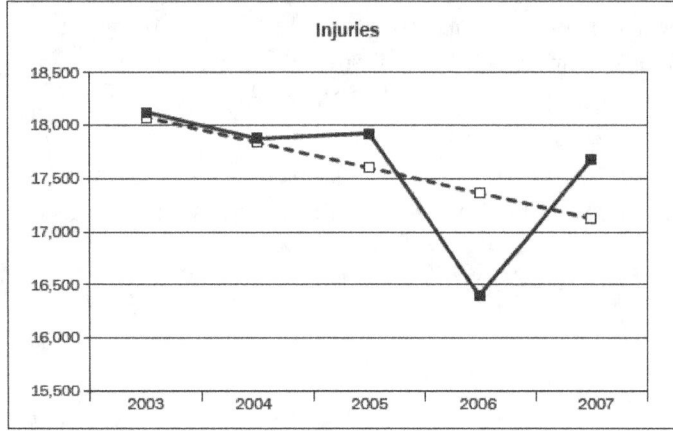

| INJURIES | |
| --- | --- |
| Year | Value |
| 2003 | 18,125 |
| 2004 | 17,875 |
| 2005 | 17,925 |
| 2006 | 16,400 |
| 2007 | 17,675 |
| 5-Year Trend (%) | -5.3% |

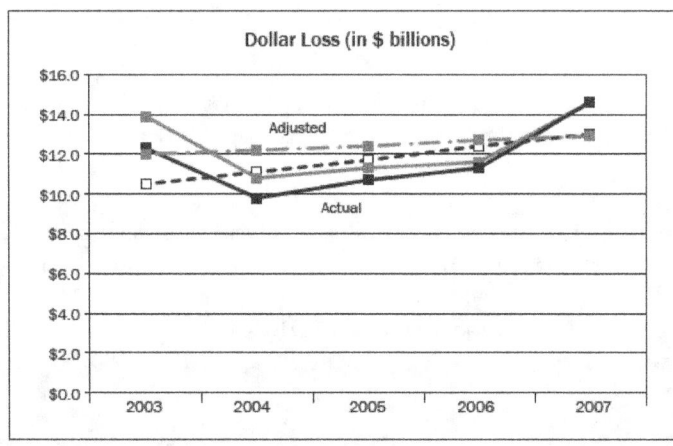

| DOLLAR LOSS ($B) | | |
| --- | --- | --- |
| Year | Actual | Adjusted to 2007 Dollars |
| 2003 | $12.3 | $13.9 |
| 2004 | $9.8 | $10.8 |
| 2005 | $10.7 | $11.3 |
| 2006 | $11.3 | $11.6 |
| 2007 | $14.6 | $14.6 |
| 5-Year Trend (%) | 23.5% | 8.1% |

Sources:   NFPA and Consumer Price Index.

In terms of dollar loss, the estimated direct value of property destroyed in fires was $15 billion for 2007, the highest value in the past 5 years.[6] The total cost of fire (direct losses, the cost of fire departments, built-in fire protection in new buildings, insurance overhead, and other annual fire protection expenditures) is considerably higher, perhaps as much as 8 to 10 times the direct losses. The direct dollar loss increased 24 percent from 2003 to 2007, with much of the increase due to inflation. Nonetheless, using constant 2007 dollars, the loss was up 8 percent over this period. The direct dollar loss was enormously high, at an average of $12 billion a year in 2007 dollars. Fire incidents increased very slightly, 1 percent, since 2003.

When USFA was established in 1974, annual fire deaths were estimated at 12,000.[7] The goal was to reduce deaths by 50 percent within 25 years; that goal was met.

On a per capita basis, despite a 1-year increase in fires in 2006, the fire problem appears less severe today than 5 years ago, partially because the population has been increasing and partially because of the overall decline in numbers of reported fires and fire casualties (Figure 3).[8] Over the 5 years, fires per million population declined 3 percent, fire deaths per million population declined 20 percent, and the injury rate declined 9 percent. In 2006, the 11 deaths per million population represented the lowest death rate in NFPA survey history. From 2003 to 2007, dollar loss per capita was up 19 percent, unadjusted. When adjusted for inflation over the 5 years, however, this loss was up only 4 percent.

## THE BROADER CONTEXT

Fires constitute a much larger problem than is generally known. Deaths and injuries from all natural disasters combined—floods, hurricanes, tornadoes, earthquakes, etc.—are just a fraction of the annual casualties from fire. For example, deaths from natural disasters average just under 200 per year, versus approximately 4,000 deaths from fires.[9,10,11]

Most fires are relatively small, and their cumulative impact is not easily recognized. Only a few fires each year have the huge dollar losses that are associated with tornados, hurricanes, or floods. The California wildfires of October 2003 resulted in over $2 billion in losses, and the California fire storm of 2007 resulted in another $1.8 billion in losses.[12] But because most of the losses from fire are spread over the nearly 1.6 million fires on average that are reported each year, the total loss is far more than the impression many people have of it from the anecdotal reporting of local fires in the media.

Fires also are an important cause of accidental deaths. For 2005, the National Safety Council (NSC) ranks fires as the sixth leading major category of nontransport accidental deaths, behind poisonings, falls, unspecified accidental factors, accidental threats to breathing, and accidental drownings.[13, 14]

---

[6] Dollar loss is rounded to the nearest $billion.

[7] NFPA changed their estimation methodology in the mid-1970s. As a result, by 1977, the estimate of fire deaths had already dropped to approximately 7,400 and rose the next year to 7,700. Nevertheless, it is fair to say that the 50-percent reduction in fire deaths was achieved.

[8] Per capita rates are determined by the number of deaths or injuries occurring to a specific population group divided by the total population for that group. This ratio is then multiplied by a common population size. For the purposes of this report, per capita rates for fire deaths and injuries are measured per 1 million persons. Per capita rates are used in the computation of relative risk.

[9] U.S. Census Bureau, *The 2008 Statistical Abstract*, http://www.census.gov/compendia/statab/2008/tables/08s0374.pdf.

[10] This average number of deaths from disasters excludes Hurricane Rita and Hurricane Katrina. When these figures are included, the yearly average is 425 deaths from natural disasters.

[11] The NCHS mortality data and the NFPA fire death estimates yield substantially different averages. Based on the NFPA survey data, fire deaths averaged 3,635 between 2003 and 2007. Because the NCHS mortality data lags, the latest 5-year average, 2002-2006 is 3,986.

[12] Stephen G. Badger, "Large Loss for 2007," NFPA Journal, November/December 2008.

[13] The category "Accidental threats to breathing" includes suffocation, accidental ingestion or inhalation of objects that obstruct the airway, and the like. Accidental drownings are not included.

[14] National Safety Council, "The odds of dying from…" http://www.nsc.org/research/odds.aspx.

## Figure 3. Fire Loss Rates (2003-2007)

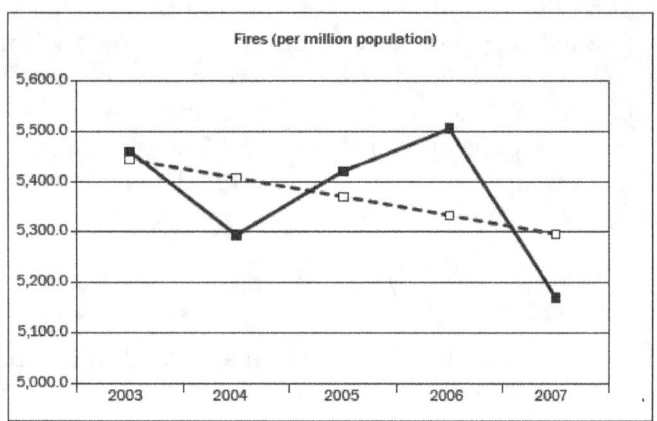

### FIRES PER MILLION POPULATION

| Year | Value |
|------|-------|
| 2003 | 5,459.8 |
| 2004 | 5,293.8 |
| 2005 | 5,420.2 |
| 2006 | 5,505.0 |
| 2007 | 5,169.4 |
| **5-Year Trend (%)** | **-2.7%** |

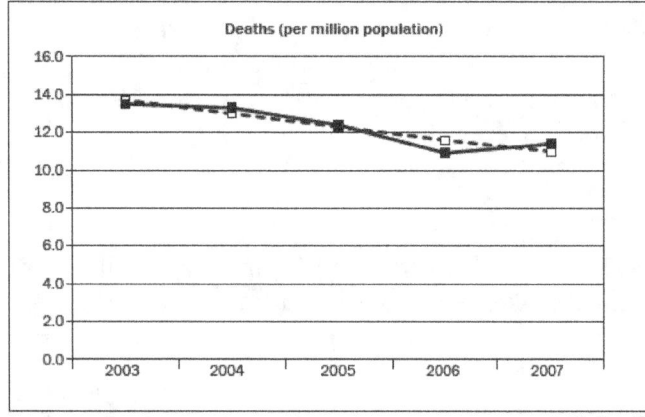

### DEATHS PER MILLION POPULATION

| Year | Value |
|------|-------|
| 2003 | 13.5 |
| 2004 | 13.3 |
| 2005 | 12.4 |
| 2006 | 10.9 |
| 2007 | 11.4 |
| **5-Year Trend (%)** | **-19.7%** |

Note: These fire death rates do not coincide with those based on the NCHS mortality data.

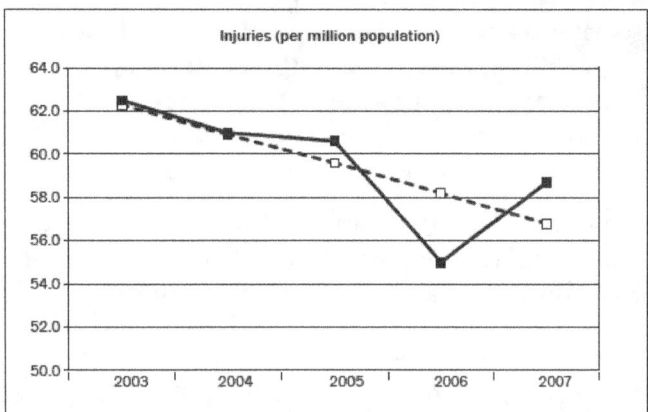

### INJURIES PER MILLION POPULATION

| Year | Value |
|------|-------|
| 2003 | 62.5 |
| 2004 | 61.0 |
| 2005 | 60.6 |
| 2006 | 55.0 |
| 2007 | 58.7 |
| **5-Year Trend (%)** | **-8.8%** |

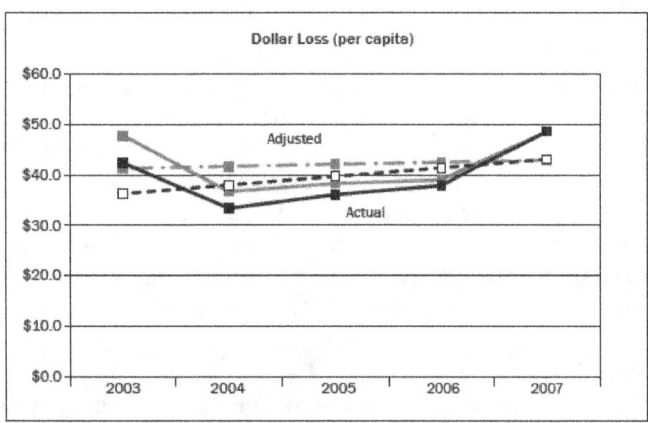

### DOLLAR LOSS PER CAPITA

| Year | Actual Value | Adjusted to 2007 Dollars |
|------|-------|-------|
| 2003 | $42.4 | $47.8 |
| 2004 | $33.4 | $36.7 |
| 2005 | $36.1 | $38.3 |
| 2006 | $37.9 | $39.0 |
| 2007 | $48.6 | $48.6 |
| **5-Year Trend (%)** | **18.5%** | **3.8%** |

Sources:    NFPA, Consumer Price Index, and U.S. Census Bureau.

Fire-related injuries to civilians and firefighters are reported with too much uncertainty to properly rank them with confidence. It is clear, however, that there were nearly 100,000 and possibly two or three times that many when injuries from unreported fires and unreported injuries from reported fires are taken into account.[15] Burn injuries are particularly tragic because of the tremendous pain and suffering they cause. Serious burns tend to cause psychological damage as well as physiological damage, and they may well involve not only the victims but also their families, friends, and fellow workers.

## U.S. Fire Deaths versus Other Nations

Although the United States no longer has one of the most severe fire problems among the industrialized Nations, it continues to experience fire death and property loss rates in excess of its sister industrialized Nations. Much progress has been made in 30 years—the death rate is less than half of what it was in the late 1970s and down 20 percent since 2003 (Figure 3). International data, however, indicate that the United States still has a fire death rate 2 to 2 1/2 times that of several European Nations, and at least 20 percent higher than many other Nations. The U.S. fire death rate, averaged for 2003 to 2005, was reported at 14.1 deaths per million population.[16] Of the 25 industrial Nations examined by the World Fire Statistics Centre, the U.S. rate is still in the upper tier—the fifth highest fire death rate out of 25 Nations. This general status has been unchanged for the past 27 years.

The declining U.S. trend in the fire death rate is not an extraordinary event; this broad declining trend applies to western European Nations and selected industrialized Nations of southeastern Asia. The United States has placed greater emphasis on fire suppression than other Nations, but these Nations tend to surpass the U.S. in practicing fire prevention. The United States would be well-served by studying and implementing international fire prevention programs that have proved effective in reducing the number of fires and deaths. The United States has excellent building technology; public buildings generally have good records. It is the combination of safety built into homes and safety behavior in homes where we fall short of some Nations. We have the technology in home sprinkler systems and knowledge of compartmentalization, but they are not widely used.

## Total Cost of Fire

The total cost of fire to society is staggering—over $182 billion per year.[17] This includes the cost of adding fire protection to buildings, the cost of paid fire departments, the equivalent cost of volunteer fire departments, the cost of insurance overhead, the direct cost of fire-related losses, the medical cost of fire injuries, and other direct and indirect costs. Even if this estimate is overstated by as much as 100 percent, the total cost of fire would range from approximately $90 to $182 billion, still enormous, and on the order of 1 to 2 percent of the gross domestic product, which was $13.8 trillion in 2007.[18] Thus, from an economic viewpoint, fire ranks as a significant national problem.

---

[15] In 2007, civilian injuries totaled 17,675 and firefighter injuries totaled 80,100, for a total of 97,775 fire-related injuries.

[16] "World Fire Statistics," Geneva Association Information Newsletter, Oct. 2008, *http://www.genevaassociation.org/PDF/WFSC/GA2008-FIRE24.pdf*. As reported, the United States has a fire death rate of 1.41 fire deaths per 100,000 population for 2003 to 2005; Singapore's rate, the lowest of the Nations studied, was .12 per 100,000 population. Austria has the lowest comparable European death rate at 0.57 per 100,000 population. Switzerland is lower still but only includes fire deaths in buildings and excludes firefighter deaths.

[17] Meade, William P., *A First Pass at Computing the Cost of Fire in a Modern Society*, The Herndon Group, Inc., February 1991. Meade estimated the cost of fire at $115 billion in that publication. The figure quoted here is adjusted for inflation to 2007 dollars to match the other loss figures quoted in this document. NFPA estimates the cost at a much higher figure, $317 billion for 2006, largely due to the cost attributed to the value of volunteer firefighter time ($119 billion) (*http://www.nfpa.org/assets/files//PDF/totalcost.pdf*).

[18] U.S. Department of Commerce, Bureau of Economic Analysis, *http://www.bea.gov/national/xls/gdplev.xls*.

# FIRE CASUALTIES BY POPULATION GROUP

The fire problem is more severe for some groups than others. People in the Southeast, males, the elderly, African-Americans, and American Indians all are at higher risk from fire than the rest of the population. These groups have remained at risk despite continuing downward trends.

## State and Regional Profiles

Fire death rates for each State for 2002 to 2006 are shown in Figure 4. An overlay plot on each State chart shows the national fire death rate. Seventeen States and the District of Columbia are consistently above the national average and 18 States are consistently below it.

The rank order of State fire deaths per million population is shown in Figure 5. States with relatively small populations may move up and down on the list from year-to-year as a result of only a few deaths; their death rate is better viewed as an average over time. For example, Alaska changed from one of the lower rates in 2004 to one of the highest in 2006. Rhode Island went from the worst fire death rate in 2003 (due to 100 deaths resulting from The Station nightclub fire) to the second best in 2006. Along with the District of Columbia, the States with the highest fire death rates in 2006 were West Virginia, Tennessee, and Kansas. The lowest were New Hampshire, Rhode Island, Massachusetts, and Hawaii.

Figure 6 shows the rank order of States in terms of the absolute number of fire deaths. Not surprisingly, large population States are at the top of the list. As in most previous years, the 10 States with the most fire deaths account for over half of the national total. Unless their fire problems are significantly reduced, the national total will be difficult to lower.

The sum of the fire deaths (3,940) is nearly 700 deaths above the estimate of 3,245 from the NFPA survey for 2006. This substantial difference may be due to the methodology and International Classification of Disease (ICD) codes used to extract fire deaths from the National Center for Health Statistics (NCHS) mortality data, or an underestimate from the extrapolation of the NFPA sample of fire departments, or a combination of both.[19,20] In most years, the comparability between the sources, while not exact, is generally considered good.

---

[19] In many years, the NCHS fire death counts and the NFPA estimates are statistically close. The 95 percent confidence intervals for the NFPA estimate of fire deaths for 2006 result in a range of 2,905 to 3,585. The NFPA fire death range is still below the NCHS fire death count for this year.

[20] For each reported death certificate in the United States, NCHS assigns codes for all reported conditions leading to death. Based on NCHS mortality data, there were 3,940 fire-related deaths in 2006. These include all deaths in which exposure to fire, fire products, or explosion was the underlying cause of death or was a contributing factor in the chain of events leading to death. This latter condition is an expanded approach to capturing fire and fire-related deaths. With this current approach, deaths where such exposures were a contributing factor (i.e., the death may not have occurred without the exposure) can be captured. Previous data and methodologies were able to capture only those deaths that directly resulted from the exposure to fire and fire products, and yielded more conservative numbers. The most conservative definition (fire and flame only, International Classification of Disease codes X00–09) yields 3,122 fire-related deaths for 2006. The codes included in this report's mortality statistics are F63.1, W39-W40, X00-X09, X75-76, X96-97, Y25-26, and Y35.1.

Chapter 2: The National Fire Problem

Figure 4.
5-Year Fire Death Rates by State
Compared to National Average

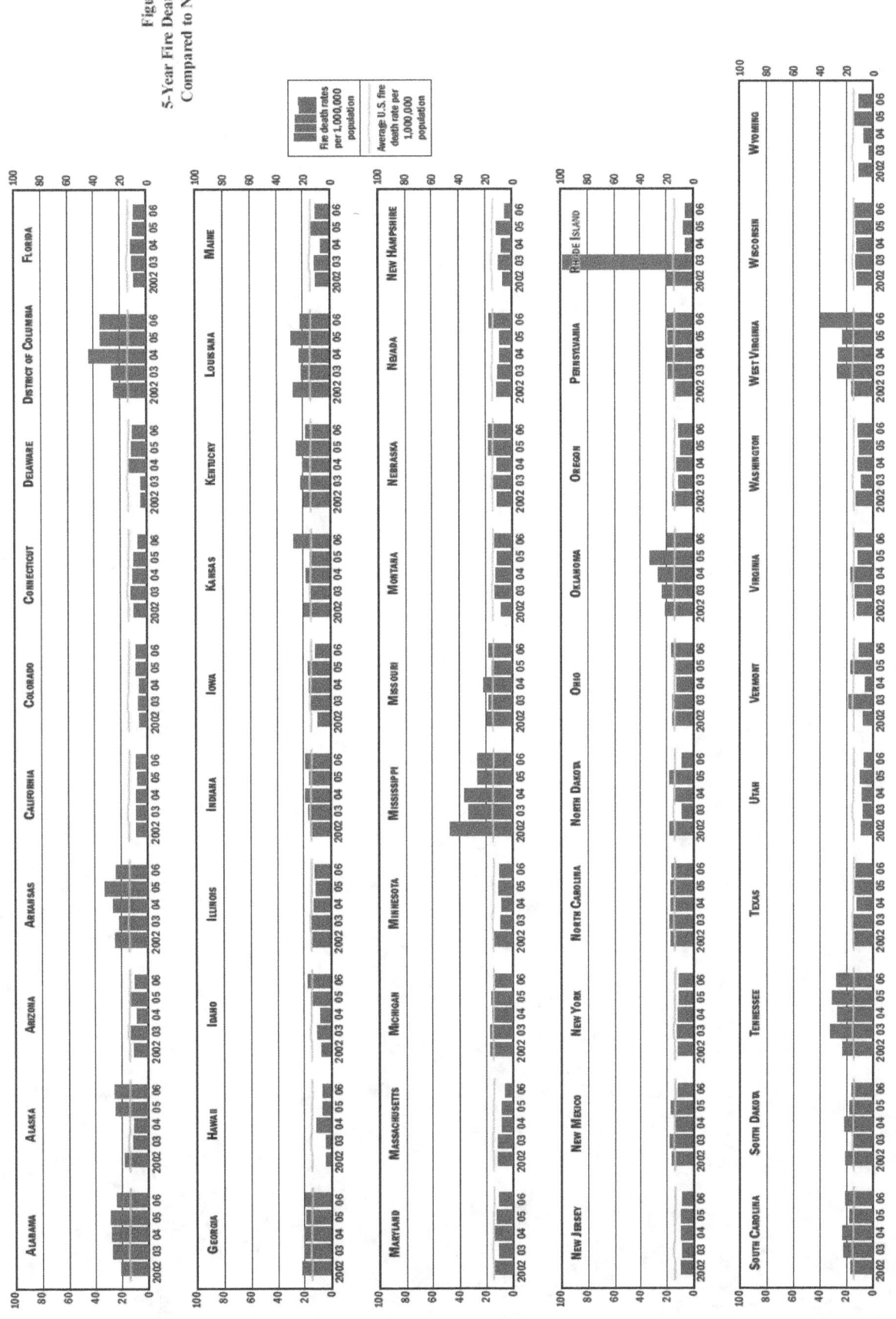

Sources: National Center for Health Statistics and U.S. Census Bureau.

## Figure 5. Rank Order of States by Civilian Fire Deaths per Million Population (2006)

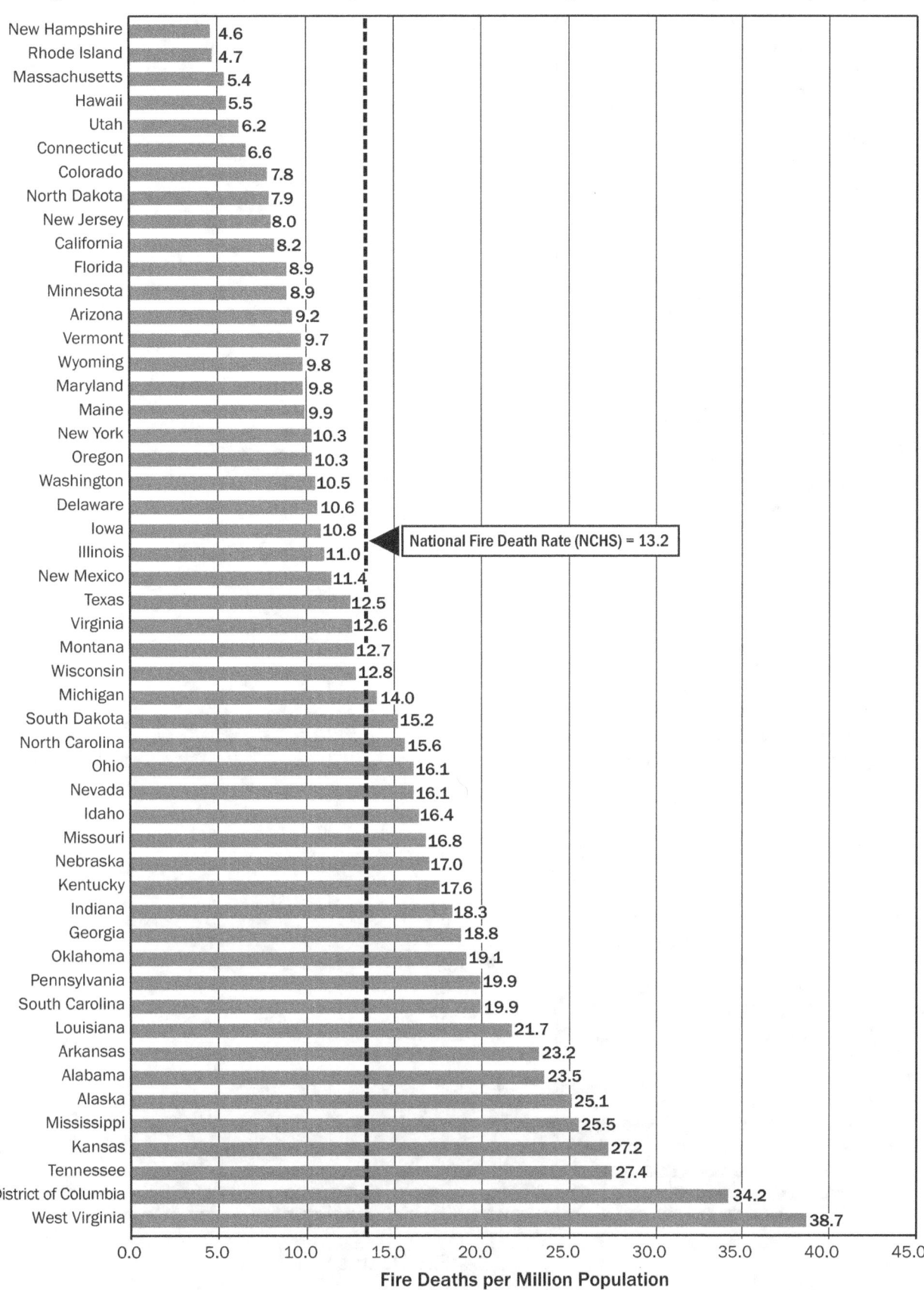

| State | Fire Deaths per Million Population |
|---|---|
| New Hampshire | 4.6 |
| Rhode Island | 4.7 |
| Massachusetts | 5.4 |
| Hawaii | 5.5 |
| Utah | 6.2 |
| Connecticut | 6.6 |
| Colorado | 7.8 |
| North Dakota | 7.9 |
| New Jersey | 8.0 |
| California | 8.2 |
| Florida | 8.9 |
| Minnesota | 8.9 |
| Arizona | 9.2 |
| Vermont | 9.7 |
| Wyoming | 9.8 |
| Maryland | 9.8 |
| Maine | 9.9 |
| New York | 10.3 |
| Oregon | 10.3 |
| Washington | 10.5 |
| Delaware | 10.6 |
| Iowa | 10.8 |
| Illinois | 11.0 |
| New Mexico | 11.4 |
| Texas | 12.5 |
| Virginia | 12.6 |
| Montana | 12.7 |
| Wisconsin | 12.8 |
| Michigan | 14.0 |
| South Dakota | 15.2 |
| North Carolina | 15.6 |
| Ohio | 16.1 |
| Nevada | 16.1 |
| Idaho | 16.4 |
| Missouri | 16.8 |
| Nebraska | 17.0 |
| Kentucky | 17.6 |
| Indiana | 18.3 |
| Georgia | 18.8 |
| Oklahoma | 19.1 |
| Pennsylvania | 19.9 |
| South Carolina | 19.9 |
| Louisiana | 21.7 |
| Arkansas | 23.2 |
| Alabama | 23.5 |
| Alaska | 25.1 |
| Mississippi | 25.5 |
| Kansas | 27.2 |
| Tennessee | 27.4 |
| District of Columbia | 34.2 |
| West Virginia | 38.7 |

National Fire Death Rate (NCHS) = 13.2

**Fire Deaths per Million Population**

Sources: National Center for Health Statistics and U.S. Census Bureau.

## Figure 6. Rank Order of States by Civilian Fire Deaths (2006)

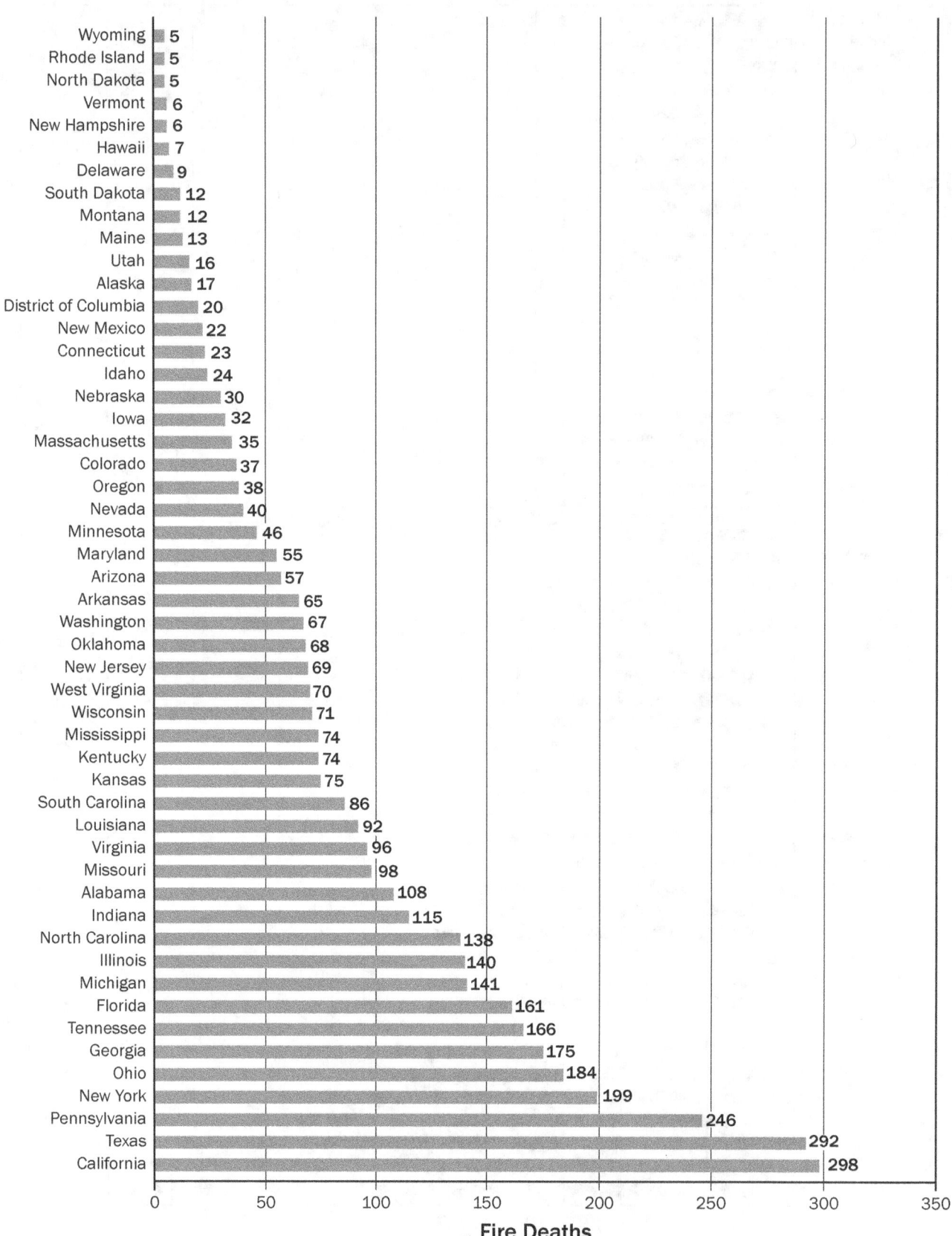

**Fire Deaths**

The Southeast of the United States continues to have the overall highest fire death rate in the Nation and one of the highest in the world. Figure 7 shows the States grouped by fire death rates for 2006. As can be seen from the map, blocks of contiguous States often have similar death rates. A special study on the commonality among these similar States might provide useful insights into the reasons for this.

**Figure 7. Fire Death Rate by State (2006)**

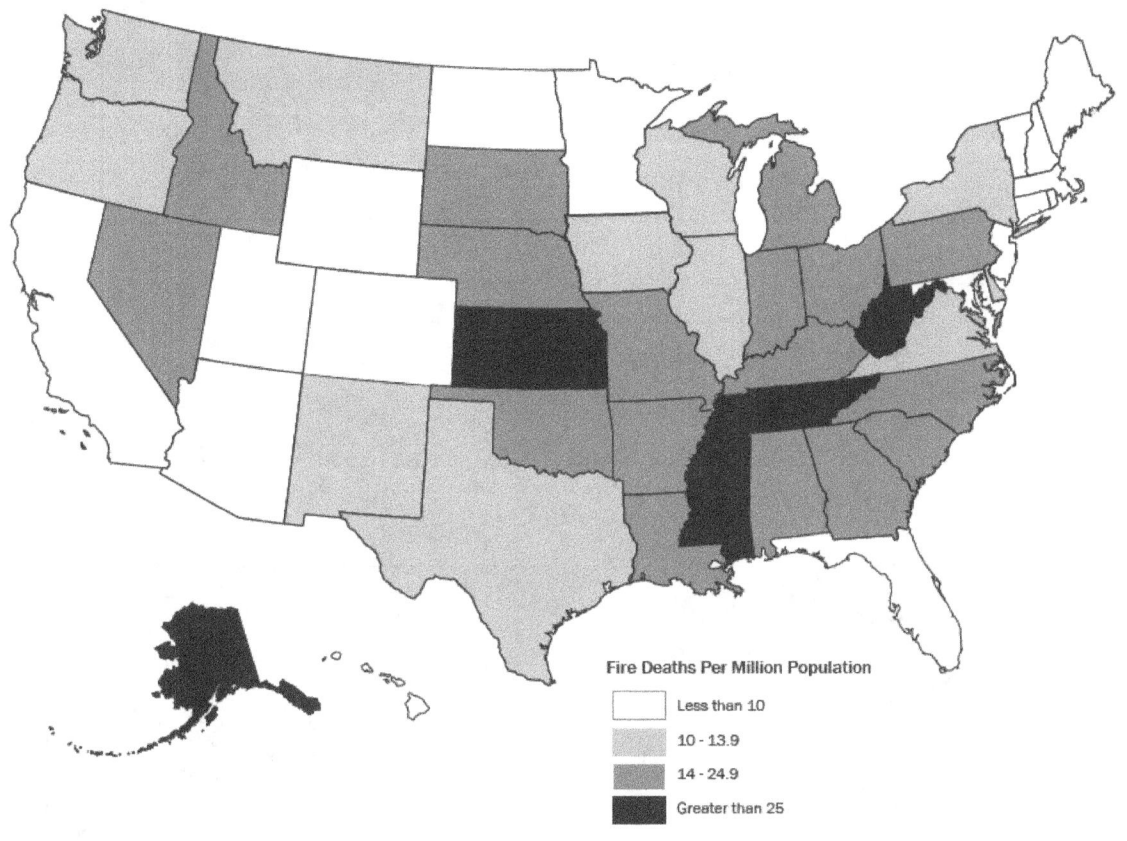

Sources:   National Center for Health Statistics and U.S. Census Bureau.

For most of the southeastern States, the fire death rates continue to decrease along with the overall U.S. rate, many still have death rates of 20 or more deaths per million population.[21] Florida continues to be the notable exception. In addition to three States from the Southeast (Mississippi, Tennessee, and West Virginia), the District of Columbia, Kansas, and Alaska were in the highest fire death rate category in 2006.

At the other extreme are the States with no shading—less than 10 deaths per million population. This death rate is in the same range as the Nations of Europe and the Far East. In the past, they tended to be States in the Southwest and West, with notable exceptions of Florida, Minnesota, and the northern New England States. In 2006, this pattern has shifted towards the New England States with fewer Western and Southwestern States and includes North Dakota along with Minnesota.  California and Florida continue to have the lowest death rates among the high-population States, as they have had for many years.

---

[21] The U.S. Census Bureau does not provide a standard definition for the Southeastern United States as a census region. The definition used here includes Alabama, Florida, Georgia, Kentucky, Louisiana, Mississippi, North Carolina, South Carolina, Tennessee, Virginia, and West Virginia as noted in the University of California-Riverside Library Glossary, http://lib.ucr.edu/depts/acquisitions/YBP%20NSP%20GLOSSARY%20EXTERNAL%20revised6-02.php.

## *Deaths*

### Gender

More men die in fires than women (Table 8). The high proportion of male-to-female fire deaths has remained remarkably steady as shown in the previous editions of *Fire in the United States*. Males have a higher fire death rate per million population than females for all age groups, with the two exceptions of the 5 to 9 age group and the 10 to 14 age group where females had a slightly higher death rate in 2006. In general, males aged 20 to 69 had twice the fire death rate as women in 2006. Males in general have fire death rates 1 1/2 to 2 times that of females (Figure 8).

### Table 8. Distribution of Fire Deaths by Gender (2006)

| Casualty Type | Males (percent) | Females (percent) |
|---|---|---|
| Deaths | 60.3 | 39.7 |

Source: National Center for Health Statistics.

### Figure 8. Rate of Fire Deaths by Age and Gender (2006)

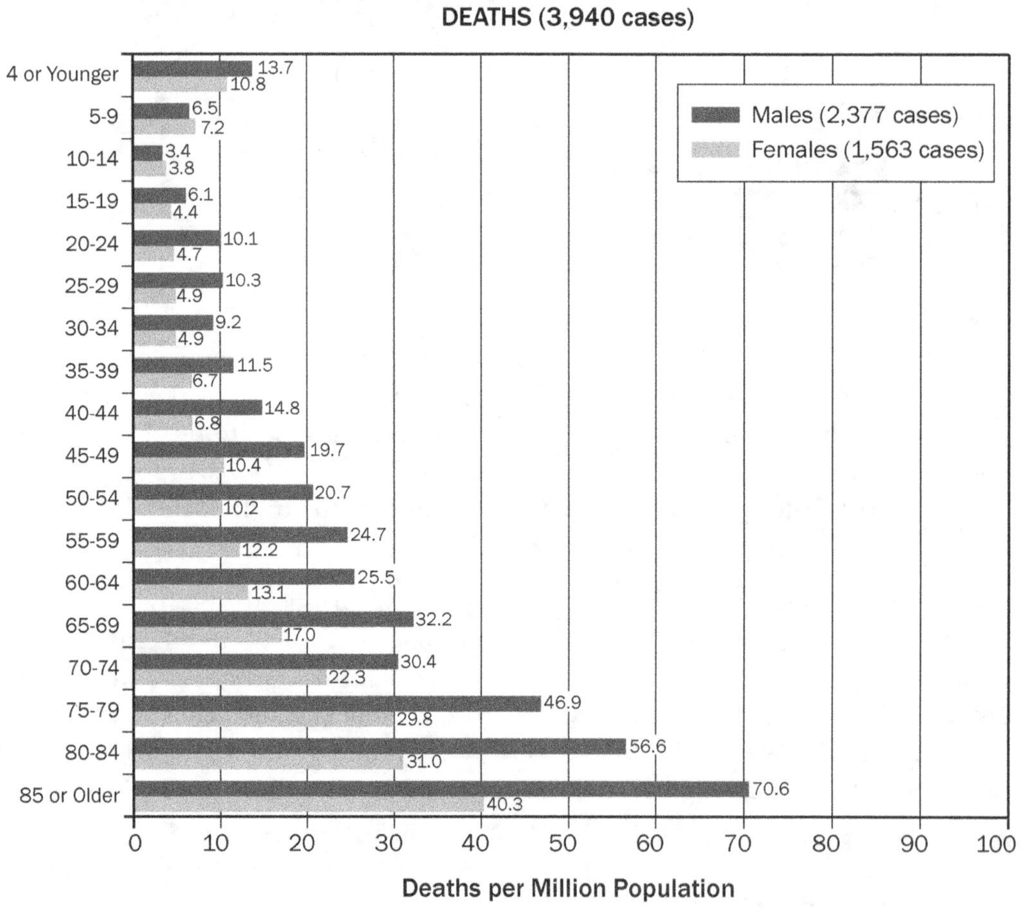

**DEATHS (3,940 cases)**

Legend: Males (2,377 cases); Females (1,563 cases)

| Age Group | Males | Females |
|---|---|---|
| 4 or Younger | 13.7 | 10.8 |
| 5-9 | 6.5 | 7.2 |
| 10-14 | 3.4 | 3.8 |
| 15-19 | 6.1 | 4.4 |
| 20-24 | 10.1 | 4.7 |
| 25-29 | 10.3 | 4.9 |
| 30-34 | 9.2 | 4.9 |
| 35-39 | 11.5 | 6.7 |
| 40-44 | 14.8 | 6.8 |
| 45-49 | 19.7 | 10.4 |
| 50-54 | 20.7 | 10.2 |
| 55-59 | 24.7 | 12.2 |
| 60-64 | 25.5 | 13.1 |
| 65-69 | 32.2 | 17.0 |
| 70-74 | 30.4 | 22.3 |
| 75-79 | 46.9 | 29.8 |
| 80-84 | 56.6 | 31.0 |
| 85 or Older | 70.6 | 40.3 |

Deaths per Million Population

Note: Data have been adjusted to account for unknown or unspecified ages.

Sources: National Center for Health Statistics and U.S. Census Bureau.

The reasons for the disparity of fire deaths between men and women are not known for certain. Suppositions include the greater likelihood of men being intoxicated and the more dangerous occupations of men (most industrial fire fatalities are males).

## Age

People over 45 have a higher fire death rate than the average population (13.2 deaths per million population), as shown in Figure 9. By age 75, the average fire death rate is higher still—nearly 3 times (or more) the national average.

The relative risk of dying in a fire for various age groups is shown in Figure 10. In 2006, the risk of fire death rose above the national average at age 45 and continued increasing for the older population groups. These profiles have remained relatively constant from year-to-year.

### Figure 9. Death Rates by Age (2006)

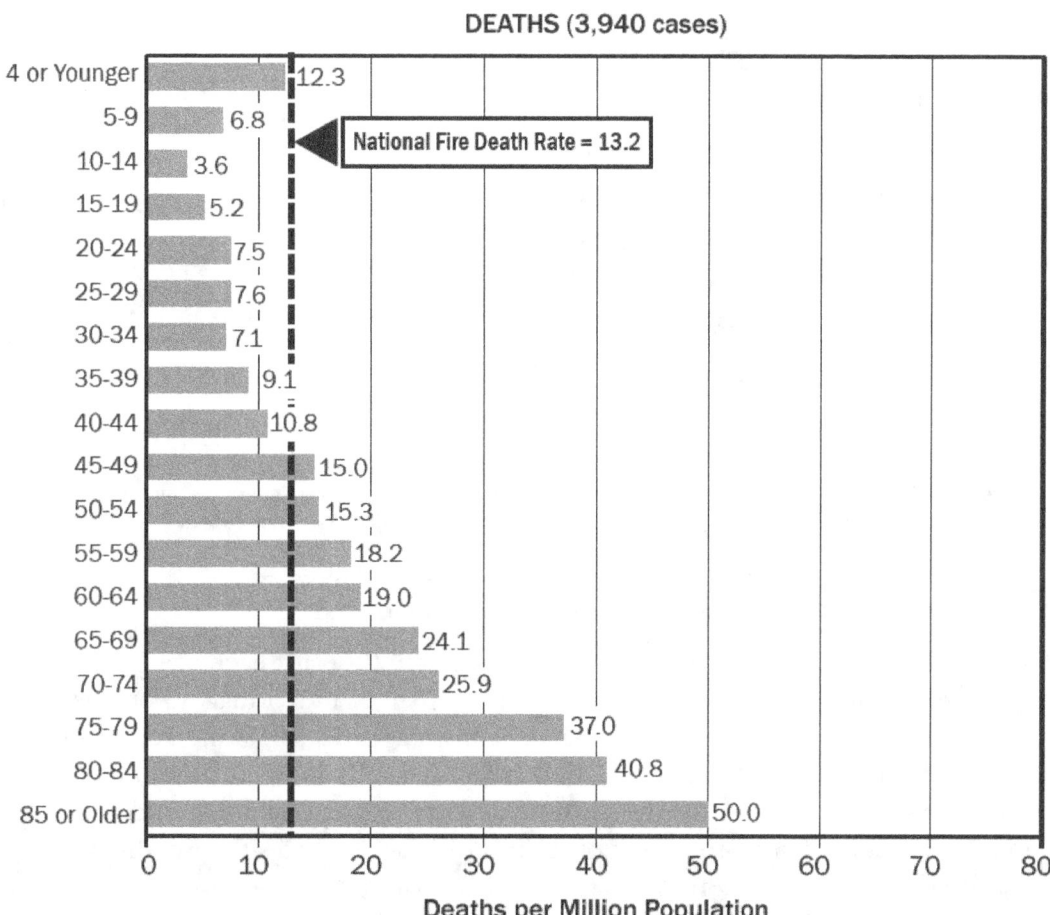

Note:     Data have been adjusted to account for unknown or unspecified ages.

Sources:   National Center for Health Statistics and U.S. Census Bureau.

## Figure 10. Relative Risk of Fire Deaths by Age (2006)

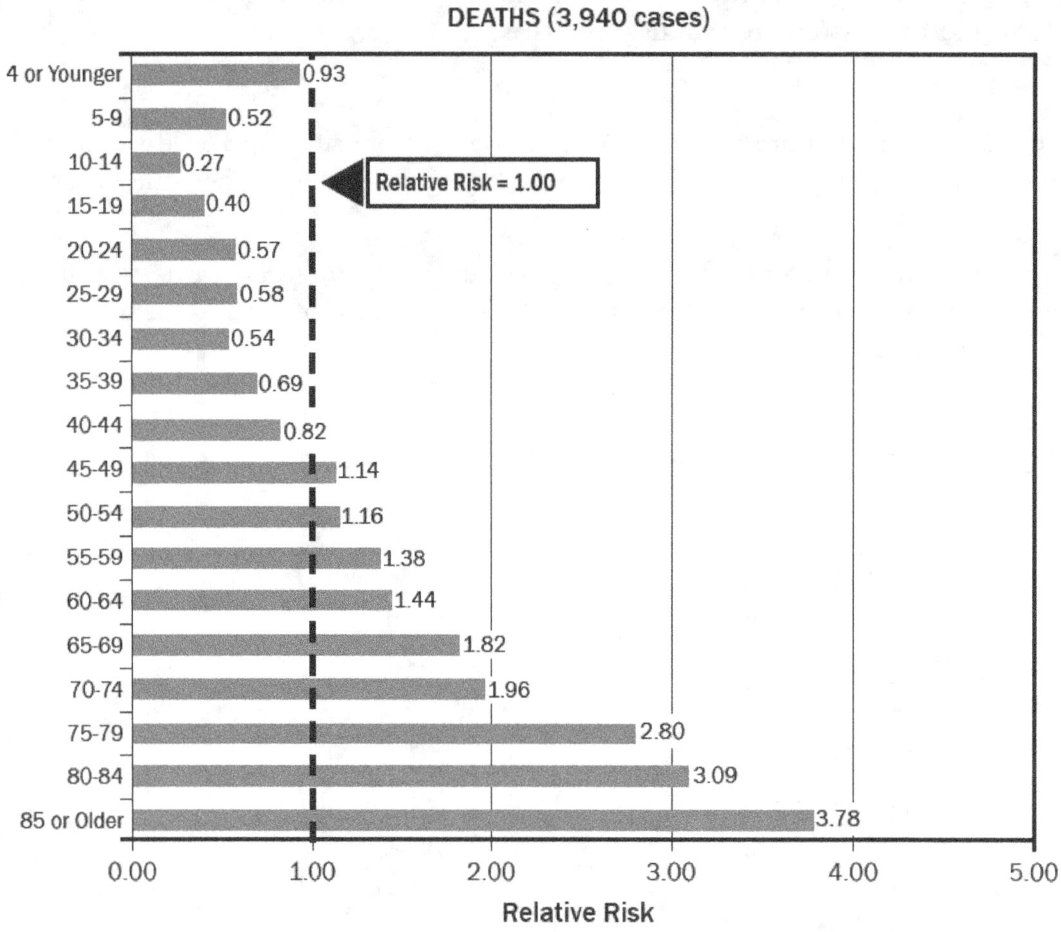

**DEATHS (3,940 cases)**

| Age Group | Relative Risk |
|-----------|---------------|
| 4 or Younger | 0.93 |
| 5-9 | 0.52 |
| 10-14 | 0.27 |
| 15-19 | 0.40 |
| 20-24 | 0.57 |
| 25-29 | 0.58 |
| 30-34 | 0.54 |
| 35-39 | 0.69 |
| 40-44 | 0.82 |
| 45-49 | 1.14 |
| 50-54 | 1.16 |
| 55-59 | 1.38 |
| 60-64 | 1.44 |
| 65-69 | 1.82 |
| 70-74 | 1.96 |
| 75-79 | 2.80 |
| 80-84 | 3.09 |
| 85 or Older | 3.78 |

Relative Risk = 1.00

**Relative Risk**

Notes:     (1) Relative risk compares the per capita rate (Figure 9) for a particular group (here, an age group) to the overall per capita rate (i.e., the general population). For the general population, the relative risk is set at 1 as indicated by the dashed lines in the above figure.
(2) Data have been adjusted to account for unknown or unspecified ages.

Sources:    National Center for Health Statistics and U.S. Census Bureau.

Figure 11 shows the percentage distribution of fire deaths in 2006 for each age group. Unlike relative risk, these percentages do not take into account the number of individuals in an age group and the distributions are somewhat different. Fire deaths peak from ages 45 to 59 and account for a combined 25 percent of the deaths. Those 4 years of age and under account for 6 percent of fire deaths, and those 65 and over comprise 32 percent of the fire deaths. These two groups, at either end of the age spectrum (the very young and older adults), represent over one-third of all fire deaths. On the other hand, nearly two-thirds of fire deaths fall in age groups that are not-so-young and not-so-old. Programs aimed only at the highest risk groups (those seen in Figure 10) will not reach the majority of potential victims.

The distribution of fire deaths by age is somewhat different for males versus females (Figure 12). Younger females (those aged 19 and under) and older females (those aged 70 and older) have a higher proportion of fire deaths than males, with elderly females over the age of 84 having nearly double the proportion of elderly males. However, the male/female proportions are reversed for those aged 20 to 69, where the proportion of fire deaths is higher for males than for females.

## Figure 11. Fire Deaths by Age (2006)

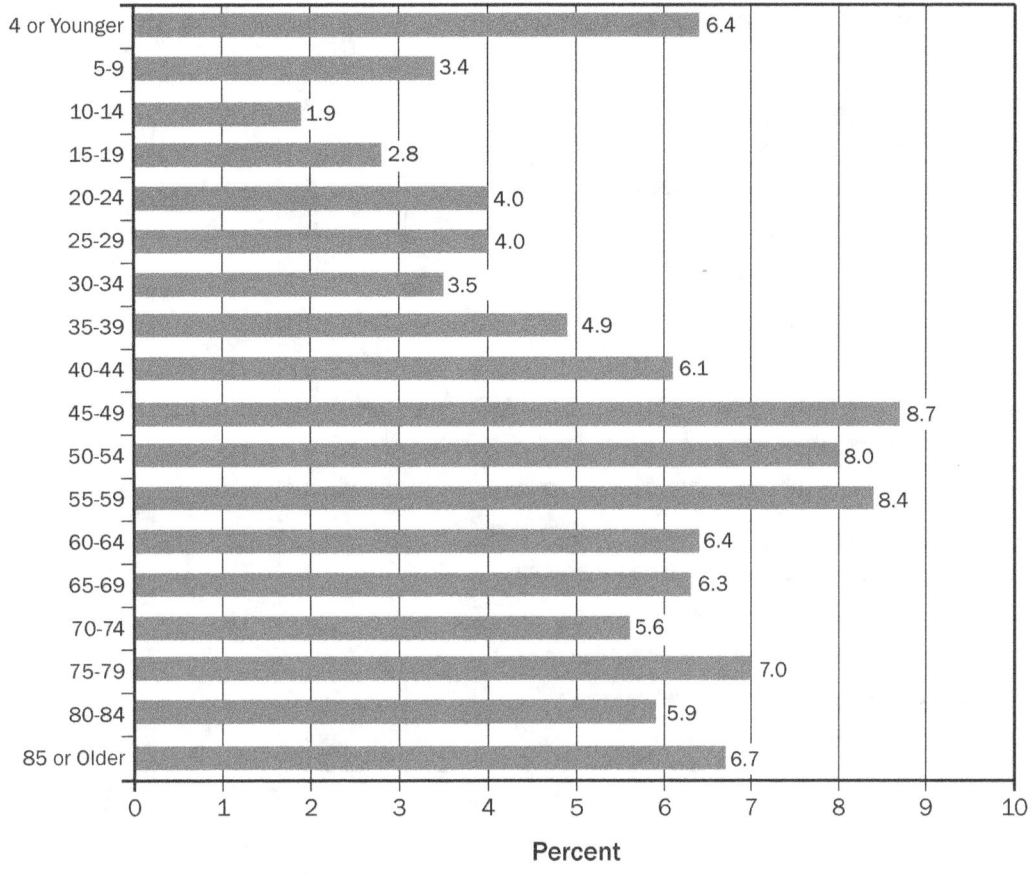

**DEATHS (3,940 cases)**

| Age | Percent |
|---|---|
| 4 or Younger | 6.4 |
| 5-9 | 3.4 |
| 10-14 | 1.9 |
| 15-19 | 2.8 |
| 20-24 | 4.0 |
| 25-29 | 4.0 |
| 30-34 | 3.5 |
| 35-39 | 4.9 |
| 40-44 | 6.1 |
| 45-49 | 8.7 |
| 50-54 | 8.0 |
| 55-59 | 8.4 |
| 60-64 | 6.4 |
| 65-69 | 6.3 |
| 70-74 | 5.6 |
| 75-79 | 7.0 |
| 80-84 | 5.9 |
| 85 or Older | 6.7 |

**Percent**

Note:     Data have been adjusted to account for unknown or unspecified ages.

Sources:   National Center for Health Statistics.

## Race

The fire problem cuts across all groups and races, rich and poor, North and South, urban and rural. But it is higher for some groups than for others.

Data on race of victims are somewhat ambiguous in a society where many people are of mixed heritage. In addition, many citizens, including firefighters, find it distasteful to report on race. On the other hand, there seems to be a higher fire problem for some groups, and it can be helpful to identify their problems for use within their own communities and by fire educators.

White males, American Indian males, and African-American males and females have higher fire death rates than the national average (Figure 13). Asian/Pacific Islanders have the lowest death rates. African-American fire death victims comprise a large and disproportionate share of total fire deaths. Although African-Americans comprise 13 percent of the population, they accounted for 22 percent of fire deaths.

## Figure 12. Fire Deaths by Age and Gender (2006)

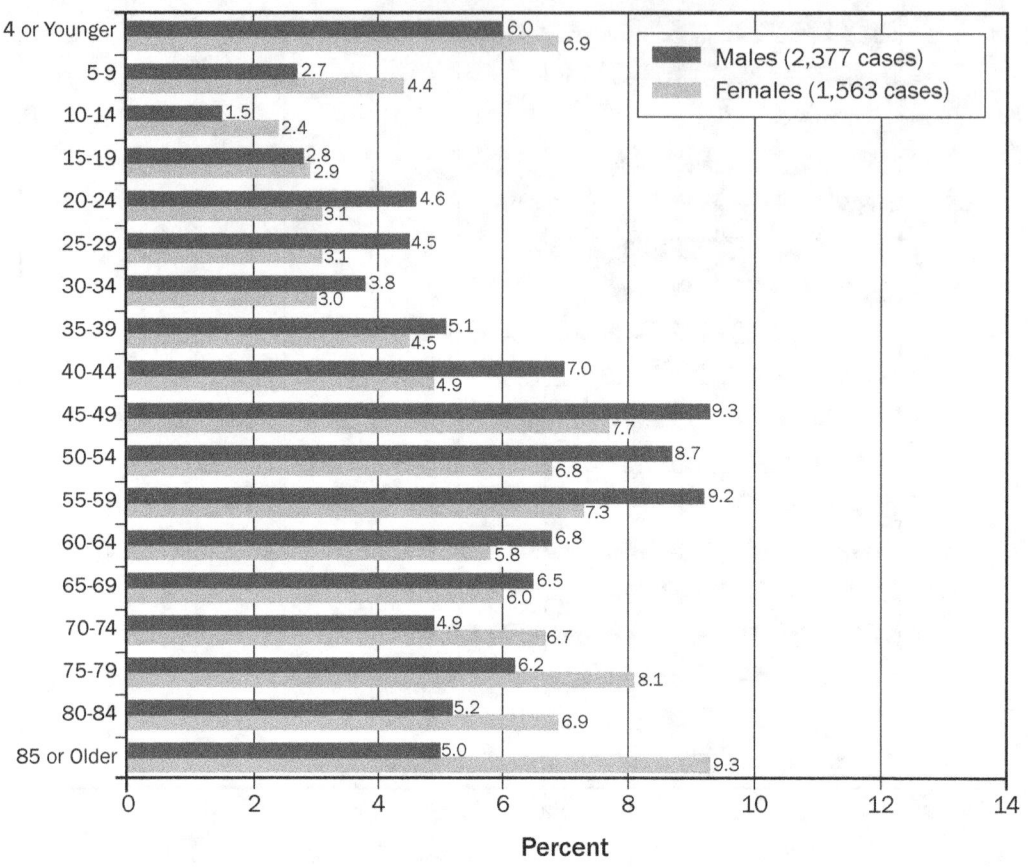

**DEATHS (3,940 cases)**

Legend: Males (2,377 cases); Females (1,563 cases)

| Age | Males | Females |
|---|---|---|
| 4 or Younger | 6.0 | 6.9 |
| 5-9 | 2.7 | 4.4 |
| 10-14 | 1.5 | 2.4 |
| 15-19 | 2.8 | 2.9 |
| 20-24 | 4.6 | 3.1 |
| 25-29 | 4.5 | 3.1 |
| 30-34 | 3.8 | 3.0 |
| 35-39 | 5.1 | 4.5 |
| 40-44 | 7.0 | 4.9 |
| 45-49 | 9.3 | 7.7 |
| 50-54 | 8.7 | 6.8 |
| 55-59 | 9.2 | 7.3 |
| 60-64 | 6.8 | 5.8 |
| 65-69 | 6.5 | 6.0 |
| 70-74 | 4.9 | 6.7 |
| 75-79 | 6.2 | 8.1 |
| 80-84 | 5.2 | 6.9 |
| 85 or Older | 5.0 | 9.3 |

Percent

Note:     Data have been adjusted to account for unknown or unspecified ages.

Source:    National Center for Health Statistics.

### Figure 13. Death Rate by Race and Gender (2006)

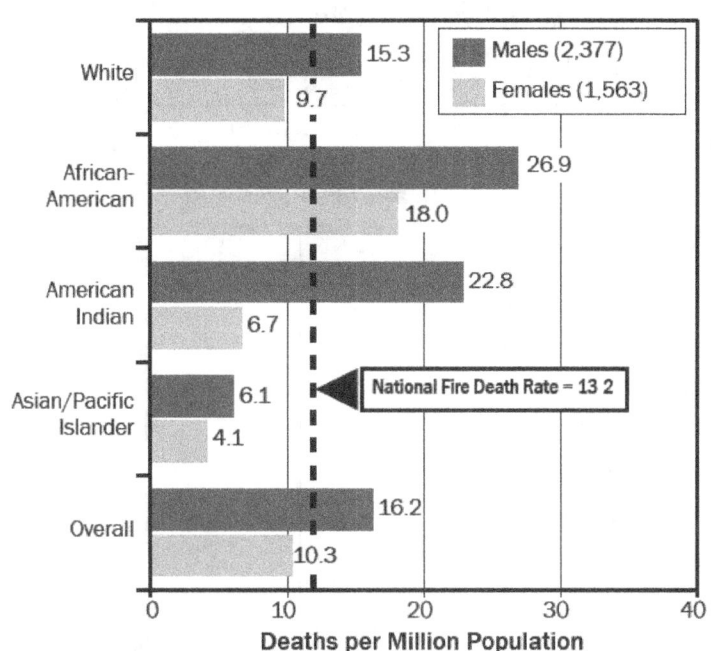

| NUMBER OF FIRE DEATHS | | | |
|---|---|---|---|
| Race | Males | Females | Total |
| White | 1,814 | 1,166 | 2,980 |
| African-American | 489 | 359 | 848 |
| American Indian | 34 | 10 | 44 |
| Asian/Pacific Islander | 40 | 28 | 68 |
| Overall | 2,377 | 1,563 | 3,940 |

Notes: (1) The overall male and female estimates include individuals with "2+ races" per the Census. The "2+ races" category accounts for approximately 1.5 percent of the population. NCHS does not include this race category.
(2) This figure uses NCHS data in the computation of the national fire death rate for data consistency within this chart. Based on NFPA fire death estimates, this rate is 10.9.

Sources: National Center for Health Statistics and U.S. Census Bureau.

## *Injuries*

### Gender

The male-to-female ratio for fire injuries is similar to that for fire deaths, except that the gender gap is slightly smaller (Table 9). Injuries per million population for males are generally 1 1/2 times the female rate, though this ratio is narrower in young children and senior citizens (Figure 14).

### Table 9. Distribution of Fire Injuries by Gender (2007)

| Casualty Type | Males (percent) | Females (percent) |
|---|---|---|
| Injuries | 58.7 | 41.3 |

Source: NFIRS.

## Figure 14. Rate of Fire Injuries by Age and Gender (2007)

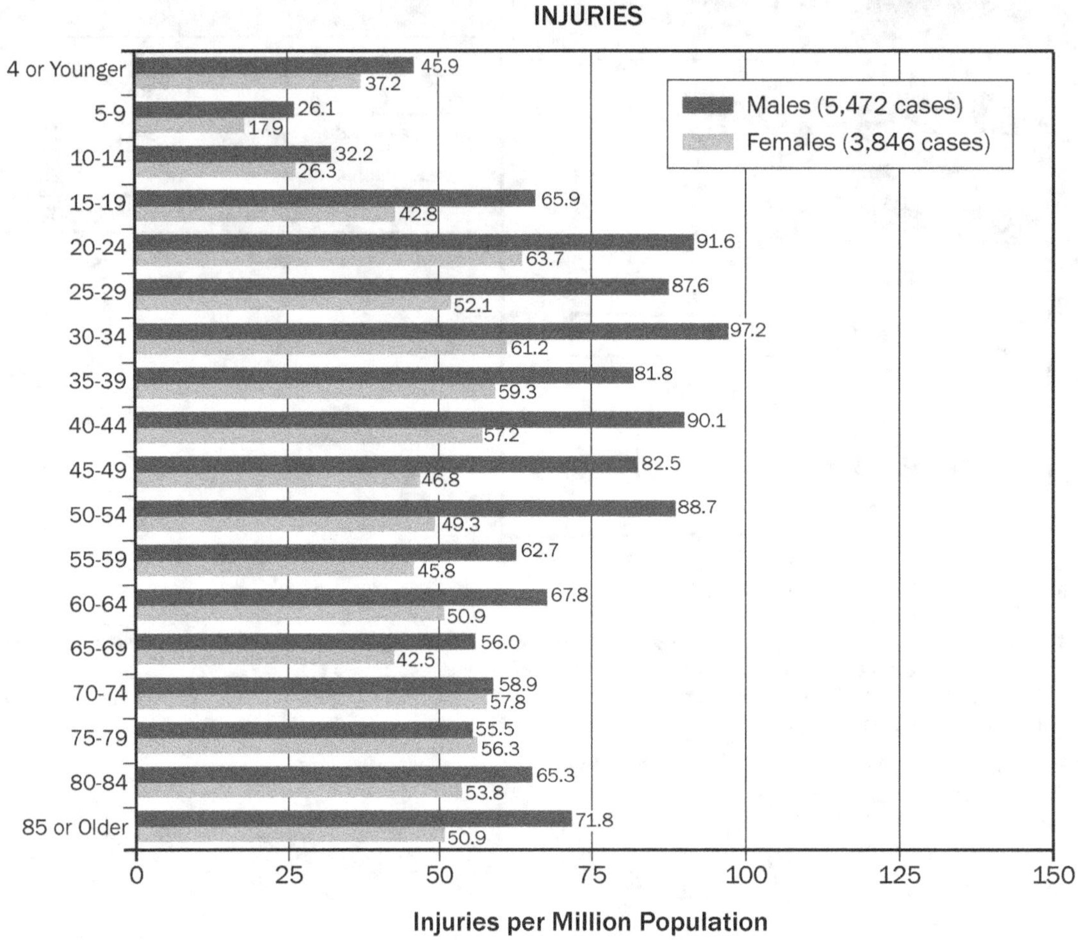

### INJURIES

| Age Group | Males | Females |
|---|---|---|
| 4 or Younger | 45.9 | 37.2 |
| 5-9 | 26.1 | 17.9 |
| 10-14 | 32.2 | 26.3 |
| 15-19 | 65.9 | 42.8 |
| 20-24 | 91.6 | 63.7 |
| 25-29 | 87.6 | 52.1 |
| 30-34 | 97.2 | 61.2 |
| 35-39 | 81.8 | 59.3 |
| 40-44 | 90.1 | 57.2 |
| 45-49 | 82.5 | 46.8 |
| 50-54 | 88.7 | 49.3 |
| 55-59 | 62.7 | 45.8 |
| 60-64 | 67.8 | 50.9 |
| 65-69 | 56.0 | 42.5 |
| 70-74 | 58.9 | 57.8 |
| 75-79 | 55.5 | 56.3 |
| 80-84 | 65.3 | 53.8 |
| 85 or Older | 71.8 | 50.9 |

Males (5,472 cases)
Females (3,846 cases)

**Injuries per Million Population**

Note:    Data have been adjusted to account for unknown or unspecified ages.

Sources:    NFIRS, NFPA, and U.S. Census Bureau.

As with fire deaths, the reasons for the disparity of fire injuries between men and women are not known for certain. However, it is known that men incur more injuries trying to extinguish the fire and rescue people than do women.

## Age

Contrary to what might be expected, the age profile for injuries is very different from that for deaths (Figure 14, Figure 15, and Figure 16). Adults aged 20 to 54 experienced the highest fire injury rates yet have some of the lowest fire death rates. Fire injury rates are below average for children and teenagers aged 19 or younger and for people aged 55 to 59, or over 64 (Figure 15). Correspondingly, the relative risk of fire injury was lowest for the younger and older age groups and highest for the mid-aged groups (Figure 16).

### Figure 15. Injury Rates by Age (2007)

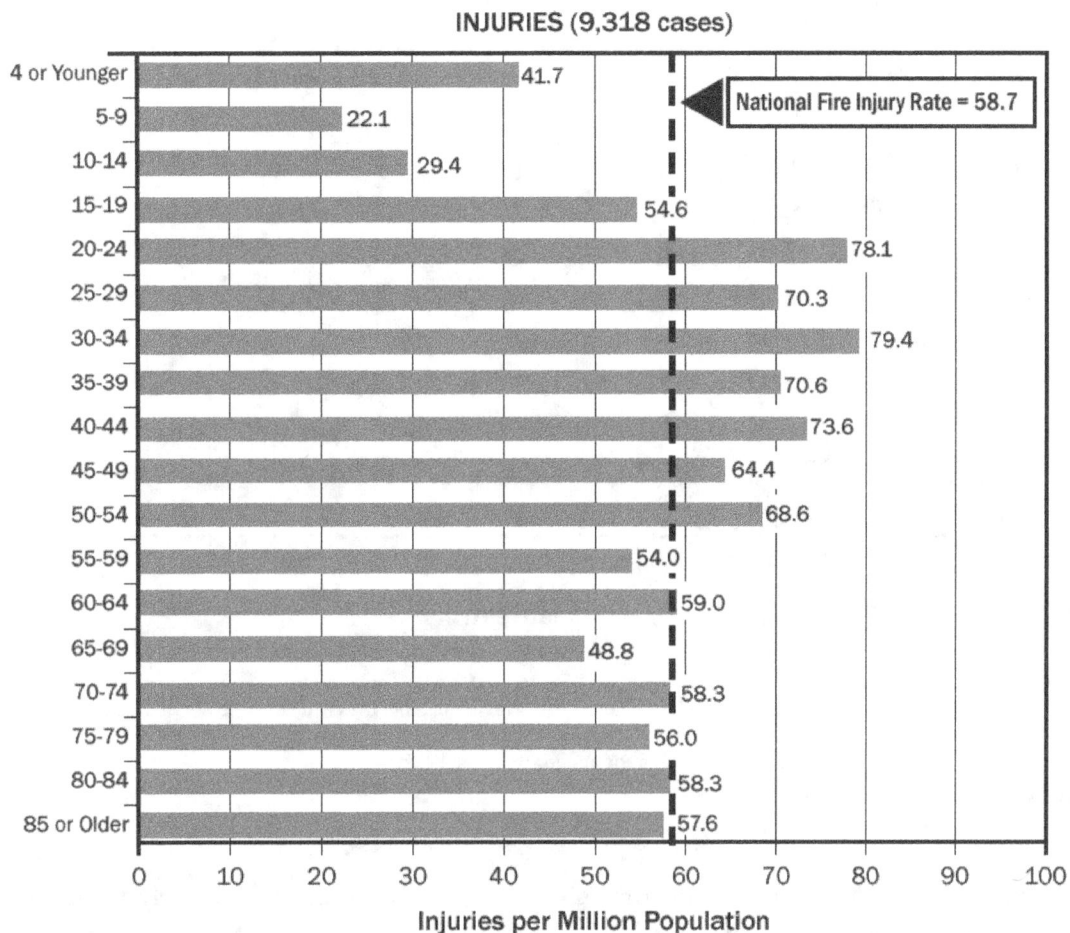

**INJURIES (9,318 cases)**

National Fire Injury Rate = 58.7

| Age | Injuries per Million Population |
|-----|-------------------------------|
| 4 or Younger | 41.7 |
| 5-9 | 22.1 |
| 10-14 | 29.4 |
| 15-19 | 54.6 |
| 20-24 | 78.1 |
| 25-29 | 70.3 |
| 30-34 | 79.4 |
| 35-39 | 70.6 |
| 40-44 | 73.6 |
| 45-49 | 64.4 |
| 50-54 | 68.6 |
| 55-59 | 54.0 |
| 60-64 | 59.0 |
| 65-69 | 48.8 |
| 70-74 | 58.3 |
| 75-79 | 56.0 |
| 80-84 | 58.3 |
| 85 or Older | 57.6 |

**Injuries per Million Population**

Note:    Data have been adjusted to account for unknown or unspecified ages.

Sources:  NFIRS, NFPA, and U.S. Census Bureau.

### Figure 16. Relative Risk of Fire Injuries by Age (2007)

**INJURIES (9,318 cases)**

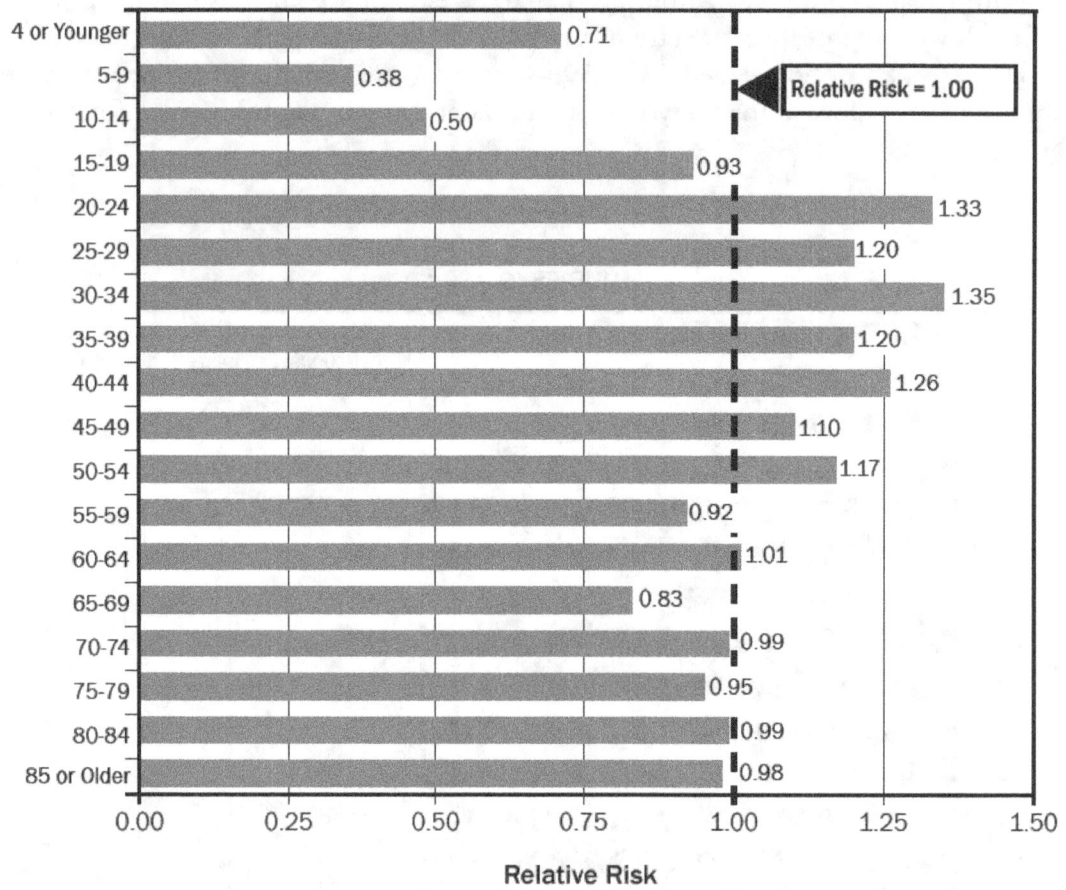

**Relative Risk**

Notes:    (1) Relative risk compares the per capita rate (Figure 15) for a particular group (here, an age group) to the overall per
          capita rate (i.e., the general population). For the general population, the relative risk is set at 1 as indicated by the
          dashed lines in the above figure.
          (2) Data have been adjusted to account for unknown or unspecified ages.

Sources:  NFIRS, NFPA, and U.S. Census Bureau.

Unlike the age distribution of fire deaths, the injury age distribution tracks closely to the relative risk profile by age for civilian fire injuries. The exception to this, however, is the elderly (Figure 17). The bulk of fire-related injuries occurs in those aged 20 to 54, and accounts for over half of the fire injuries in 2007. The young, under age 10, account for 7 percent of fire injuries; older adults over age 64 account for 12 percent. Although older adults have an average level of fire injury risk, there are fewer of them in the total population. If their risk continues to be the same, we could expect more and more elderly fire injuries and deaths as the older adult proportion of the population increases. In the meantime, the focus for injury prevention should be on adults 20 to 54. It is believed that males in this age group are greater risk takers during fires, resulting in a higher proportion of injuries.

## Figure 17. Fire Injuries by Age (2007)

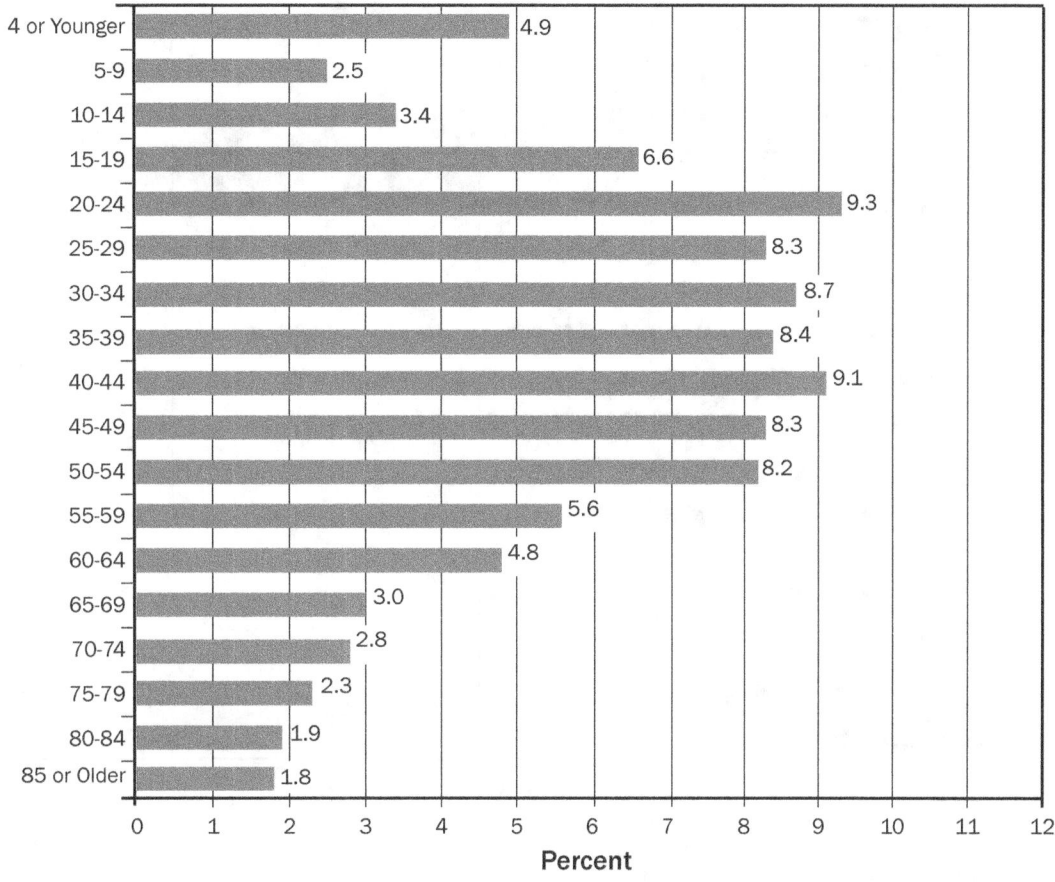

INJURIES (9,318 cases)

| Age | Percent |
|-----|---------|
| 4 or Younger | 4.9 |
| 5-9 | 2.5 |
| 10-14 | 3.4 |
| 15-19 | 6.6 |
| 20-24 | 9.3 |
| 25-29 | 8.3 |
| 30-34 | 8.7 |
| 35-39 | 8.4 |
| 40-44 | 9.1 |
| 45-49 | 8.3 |
| 50-54 | 8.2 |
| 55-59 | 5.6 |
| 60-64 | 4.8 |
| 65-69 | 3.0 |
| 70-74 | 2.8 |
| 75-79 | 2.3 |
| 80-84 | 1.9 |
| 85 or Older | 1.8 |

Percent

Note:   Data have been adjusted to account for unknown or unspecified ages.

Source:  NFIRS.

The distribution of fire injuries by age is somewhat different for males versus females (Figure 18). Males aged 15 to 54 tend to have a slightly higher proportion of injuries, while young and older females have more injuries than males when they are below the age of 15 and above the age of 54. Notably, older adult females have twice the proportion of fire injuries than older males.

**Figure 18. Fire Injuries by Age and Gender (2007)**

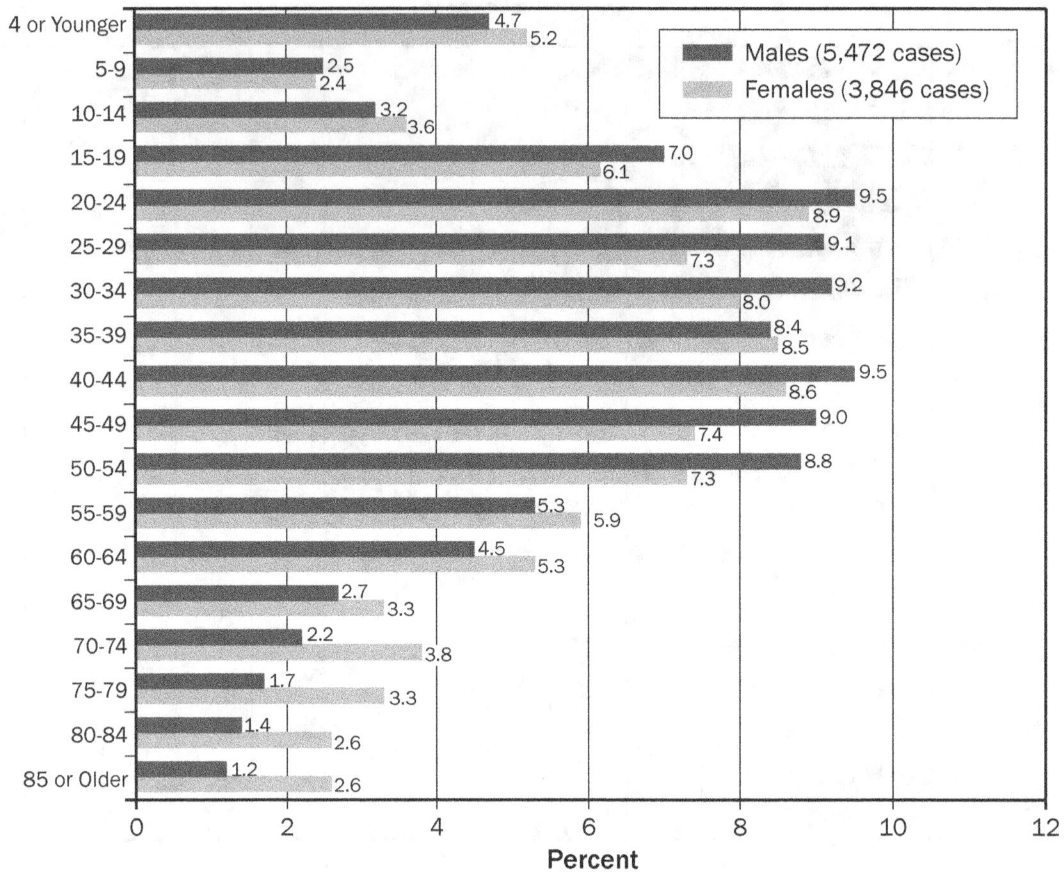

Note:    Data have been adjusted to account for unknown or unspecified ages.

Source:   NFIRS.

# KINDS OF PROPERTIES WHERE FIRES OCCUR

This section describes the proportions of the fire problem by property type: residential structures, nonresidential structures, vehicles, outside properties, and other or unknown properties.[22,23]

## *Property Types*

Over the years, there has been little change in the proportion of fires, deaths, injuries, and dollar loss by the type of property involved. In terms of numbers of fires, the largest category continues to be outside fires (44 percent)—in fields, vacant lots, trash, etc. (Figure 19). While there are many of these fires, they are not the source of most fire damage. Residential and nonresidential structure fires together comprise 35 percent of all fires, with residential structure fires outnumbering nonresidential structure fires by over three to one. What may surprise some is the large number of vehicle fires. In fact, nearly one out of every seven fires to which fire departments respond involves a vehicle.

---

[22] The percentage of fire deaths in the major property types differs somewhat between NFIRS and the NFPA annual survey. These differences are discussed in Appendix A.

[23] "Other" fires are those where the incident type is not classified or are outside gas or vapor combustion incidents.

By far, the largest percentage of deaths, 76 percent in 2007, occurs on residential properties, with the majority of these in one- and two-family dwellings. Vehicles accounted for the second largest percentage of fire deaths at 17 percent. Great attention is given to large, multiple-death fires in public places such as hotels, nightclubs, and office buildings. But the major attention-getting fires that kill 10 or more people are few in number and have constituted only a small portion of overall fire deaths. Firefighters generally are doing a good job in protecting public properties in this country. Furthermore, these properties generally are required by local codes to have built-in fire suppression systems. The area with the largest problem is where it is least suspected—in people's homes. Prevention efforts continue to be focused on home fire safety.

**Figure 19. Fire and Fire Losses by General Property Type (2007)**

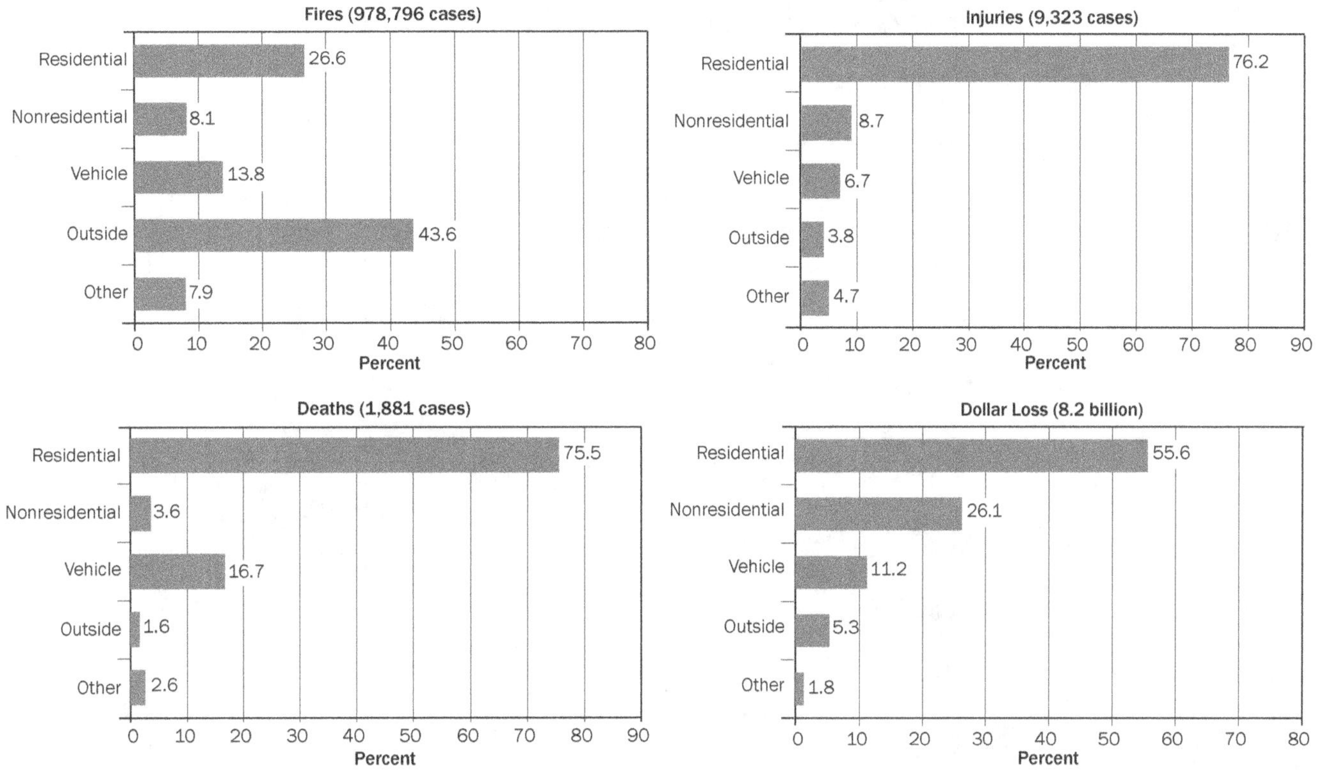

Source: NFIRS.

Only 4 percent of the 2007 fire deaths occurred in commercial and public properties. Outside and other miscellaneous fires, including wildfires, were also a small factor (4 percent combined) in fire deaths.

As Figure 19 shows, the picture is generally similar for fire injuries, with 76 percent of all injuries occurring in residences. Fire injuries are distributed to the other property types as follows: nonresidential properties, 9 percent; vehicles, 7 percent; and outside and other fires, 9 percent.

The picture changes somewhat for dollar loss. While residential structures are the leading property for dollar loss, nonresidential structures play a considerable role. These two property types account for 82 percent of all dollar loss. The proportion of dollar loss from outside fires may be understated because the destruction of trees, grass, etc., is often given zero value in fire reports if it is not commercial cropland or timber.

As a final observation on property types, structures (residential and nonresidential) account for 35 percent of fires, but they account for 79 percent of fire deaths, 85 percent of injuries, and 82 percent of dollar loss.

## Losses

Figure 20 shows casualties and losses per fire. These indicators represent the severity of fires but are somewhat ambiguous because they can increase if there are more casualties or damage per fire (the numerators) or if fewer minor fires are reported (the denominators).

Residential fires have the highest number of deaths and injuries per 1,000 fires—another important reason for prevention programs to focus on home fire safety. Nonresidential structure fires have the highest dollar loss per fire.

**Figure 20. Fire Casualties and Dollar Loss per Fire by General Property Type (2007)**

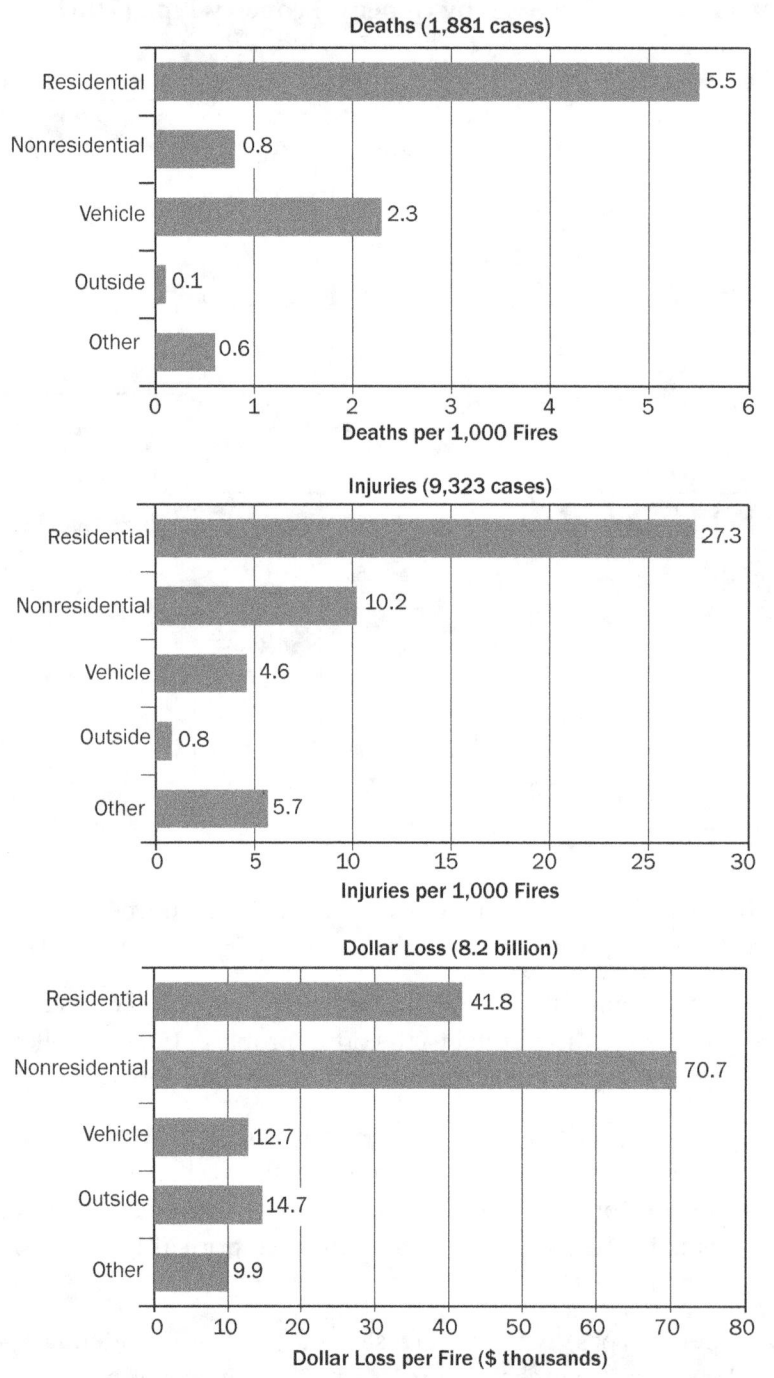

Note:    Dollar loss per fire is computed for fires causing dollar loss.

Source:  NFIRS.

# CAUSES OF FIRES AND FIRE LOSSES

The sections below discuss the fire cause profiles for 2007, by property type, of the major causes of fires and fires that caused losses: fatal fires, fires causing injuries, and fires causing dollar loss.[24]

## *Residential Structures*

Figure 21 shows the fire cause profiles for residential structures. At 40 percent, cooking is by far the leading cause of fires. The second leading cause of fires is heating, accounting for 14 percent of fires. These percentages (and those that follow) are adjusted, which proportionally spreads the incidents with unknown causes over the other 15 cause categories. The two leading causes of fatal fires—fires that result in civilian deaths—are smoking at 18 percent and other unintentional or careless actions at 14 percent. The leading cause of fires that results in injuries is cooking (26 percent), followed by other unintentional or careless actions (11 percent) and open flame (such as candles, lighters, and matches) (11 percent). Cooking is by far the leading cause of fires causing dollar loss at 20 percent.

## *Nonresidential Structures*

Cooking fires are the leading cause of nonresidential structure fires, causing nearly one quarter (23 percent) of all nonresidential structure fires (Figure 22). Intentionally set fires are the second leading cause of nonresidential structure fires (12 percent) and electrical malfunctions follow closely as the third leading cause (10 percent).

The leading cause of fatal fires in nonresidential structures is intentionally set fires, at 27 percent. The next leading causes are other types of unintentionally set fires, or carelessness (19 percent); other equipment (14 percent); and smoking (11 percent).

Interestingly, while cooking fires are by far the leading cause of nonresidential structure fires, they resulted in no fatal fires in 2007. They are, however, the leading cause of fire-related injuries (13 percent).

Intentionally set fires and electrical malfunctions are the two leading causes of fires causing dollar loss, both at 13 percent. Notably, intentionally set fires are the second leading cause of nonresidential structure fires, and are the leading causes of both fatal fires and dollar loss in nonresidential structures.

---

[24] In principle, it is the cause of the fire which results in deaths and injuries that should be analyzed, not numbers of deaths and injuries associated with fire causes.

# Figure 21. Causes of Residential Structure Fires and Fire Losses (2007)

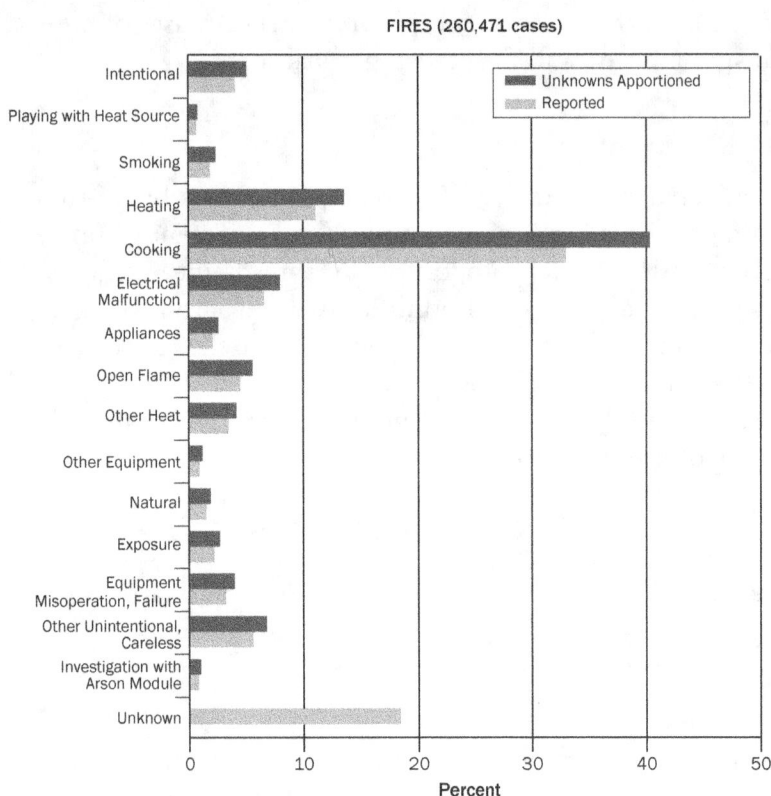

FIRES (260,471 cases)

| Cause | Reported | Unknowns Apportioned |
|---|---|---|
| Intentional | 4.1 | 5.1 |
| Playing with Heat Source | 0.7 | 0.8 |
| Smoking | 1.9 | 2.4 |
| Heating | 11.1 | 13.6 |
| Cooking | 32.9 | 40.3 |
| Electrical Malfunction | 6.6 | 8.0 |
| Appliances | 2.1 | 2.6 |
| Open Flame | 4.5 | 5.6 |
| Other Heat | 3.5 | 4.2 |
| Other Equipment | 0.9 | 1.2 |
| Natural | 1.5 | 1.9 |
| Exposure | 2.2 | 2.7 |
| Equipment Misoperation, Failure | 3.2 | 4.0 |
| Other Unintentional, Careless | 5.6 | 6.8 |
| Investigation with Arson Module | 0.8 | 1.0 |
| Unknown | 18.4 | 0.0 |

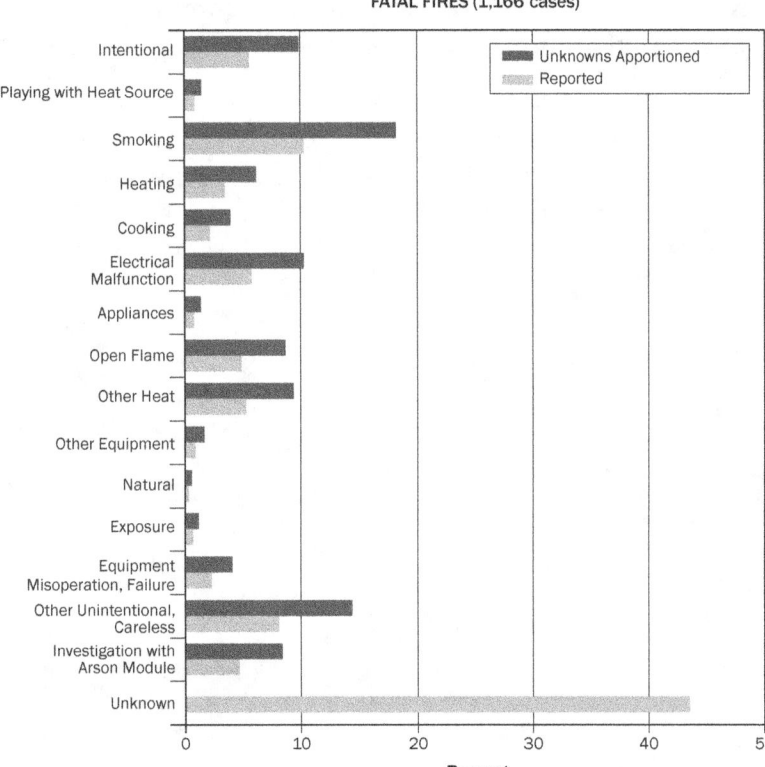

FATAL FIRES (1,166 cases)

| Cause | Reported | Unknowns Apportioned |
|---|---|---|
| Intentional | 5.6 | 9.9 |
| Playing with Heat Source | 0.9 | 1.5 |
| Smoking | 10.3 | 18.2 |
| Heating | 3.5 | 6.2 |
| Cooking | 2.2 | 4.0 |
| Electrical Malfunction | 5.8 | 10.3 |
| Appliances | 0.8 | 1.4 |
| Open Flame | 4.9 | 8.7 |
| Other Heat | 5.3 | 9.4 |
| Other Equipment | 0.9 | 1.7 |
| Natural | 0.3 | 0.6 |
| Exposure | 0.7 | 1.2 |
| Equipment Misoperation, Failure | 2.3 | 4.1 |
| Other Unintentional, Careless | 8.1 | 14.4 |
| Investigation with Arson Module | 4.7 | 8.4 |
| Unknown | 43.6 | 0.0 |

continued on next page

## Figure 21. Causes of Residential Structure Fires and Fire Losses (2007)—Continued

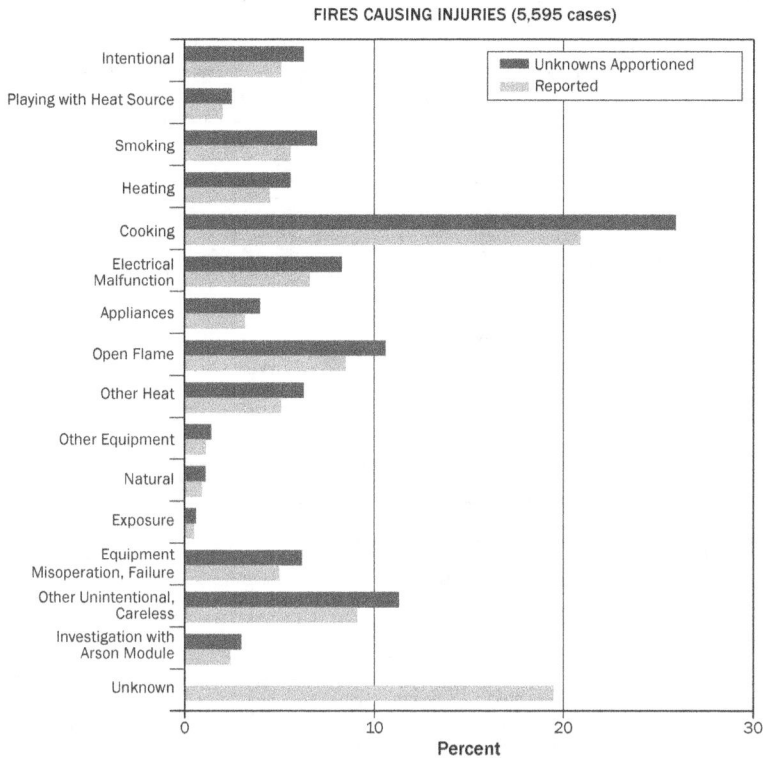

FIRES CAUSING INJURIES (5,595 cases)

| Cause | Reported | Unknowns Apportioned |
|---|---|---|
| Intentional | 5.1 | 6.3 |
| Playing with Heat Source | 2.0 | 2.5 |
| Smoking | 5.6 | 7.0 |
| Heating | 4.5 | 5.6 |
| Cooking | 20.9 | 25.9 |
| Electrical Malfunction | 6.6 | 8.3 |
| Appliances | 3.2 | 4.0 |
| Open Flame | 8.5 | 10.6 |
| Other Heat | 5.1 | 6.3 |
| Other Equipment | 1.1 | 1.4 |
| Natural | 0.9 | 1.1 |
| Exposure | 0.5 | 0.6 |
| Equipment Misoperation, Failure | 5.0 | 6.2 |
| Other Unintentional, Careless | 9.1 | 11.3 |
| Investigation with Arson Module | 2.4 | 3.0 |
| Unknown | 19.5 | 0.0 |

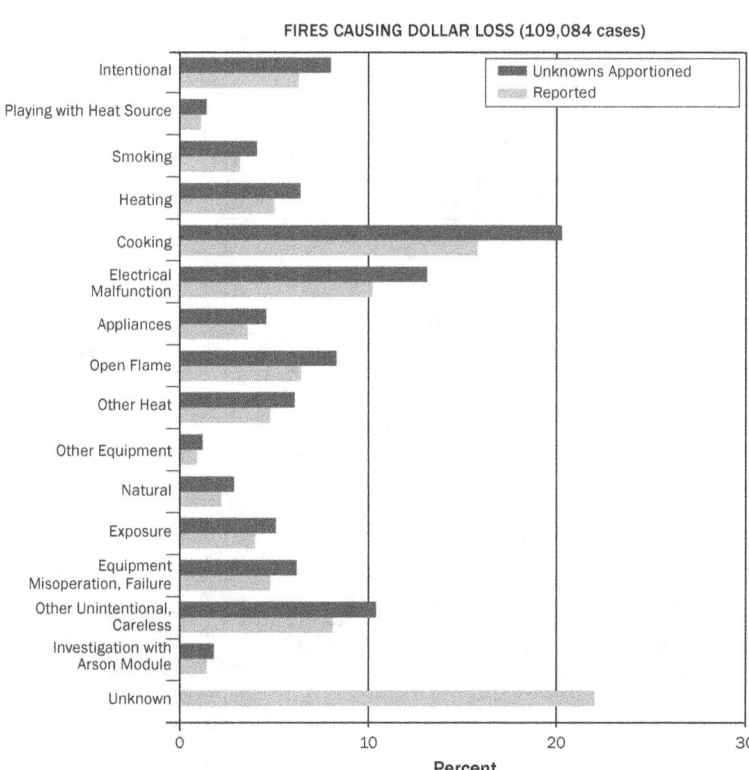

FIRES CAUSING DOLLAR LOSS (109,084 cases)

| Cause | Reported | Unknowns Apportioned |
|---|---|---|
| Intentional | 6.3 | 8.0 |
| Playing with Heat Source | 1.1 | 1.4 |
| Smoking | 3.2 | 4.1 |
| Heating | 5.0 | 6.4 |
| Cooking | 15.8 | 20.3 |
| Electrical Malfunction | 10.2 | 13.1 |
| Appliances | 3.6 | 4.6 |
| Open Flame | 6.4 | 8.3 |
| Other Heat | 4.8 | 6.1 |
| Other Equipment | 0.9 | 1.2 |
| Natural | 2.2 | 2.9 |
| Exposure | 4.0 | 5.1 |
| Equipment Misoperation, Failure | 4.8 | 6.2 |
| Other Unintentional, Careless | 8.1 | 10.4 |
| Investigation with Arson Module | 1.4 | 1.8 |
| Unknown | 22.0 | 0.0 |

Source: NFIRS.

# Figure 22. Causes of Nonresidential Structure Fires and Fire Losses (2007)

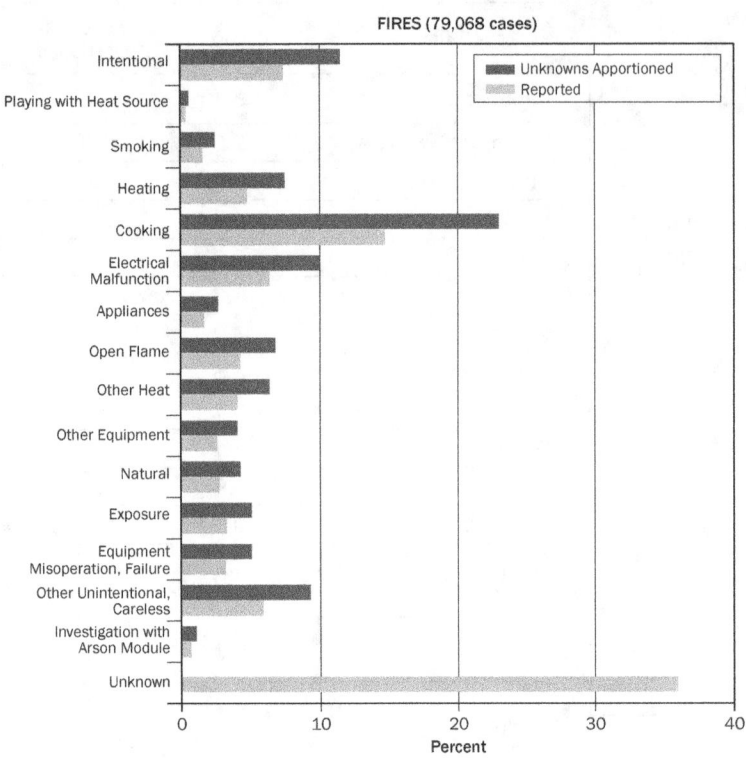

FIRES (79,068 cases)

| Cause | Reported | Unknowns Apportioned |
|---|---|---|
| Intentional | 7.4 | 11.5 |
| Playing with Heat Source | 0.4 | 0.6 |
| Smoking | 1.6 | 2.5 |
| Heating | 4.8 | 7.5 |
| Cooking | 14.7 | 23.0 |
| Electrical Malfunction | 6.4 | 10.0 |
| Appliances | 1.7 | 2.7 |
| Open Flame | 4.3 | 6.8 |
| Other Heat | 4.1 | 6.4 |
| Other Equipment | 2.6 | 4.1 |
| Natural | 2.8 | 4.3 |
| Exposure | 3.3 | 5.1 |
| Equipment Misoperation, Failure | 3.2 | 5.1 |
| Other Unintentional, Careless | 5.9 | 9.3 |
| Investigation with Arson Module | 0.7 | 1.1 |
| Unknown | 36.0 | 0.0 |

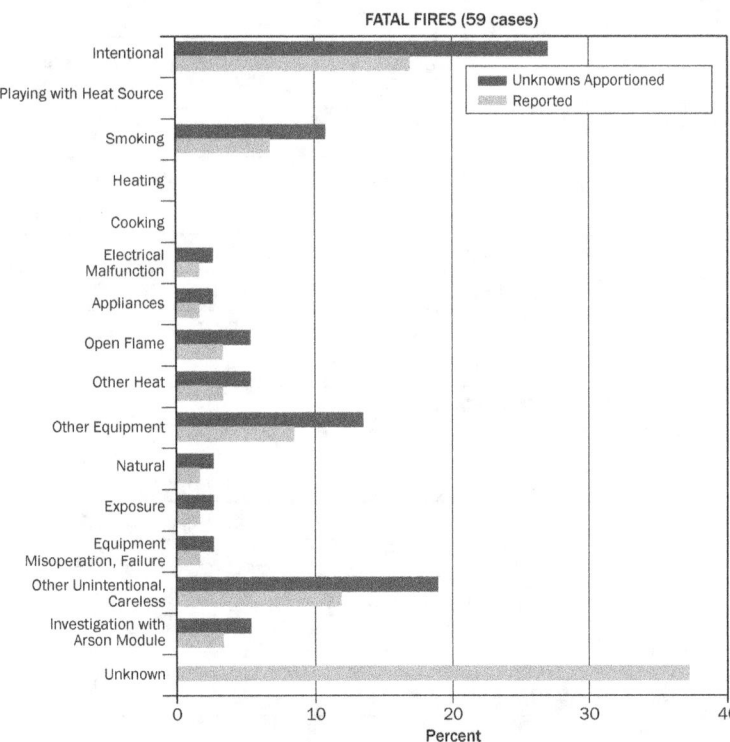

FATAL FIRES (59 cases)

| Cause | Reported | Unknowns Apportioned |
|---|---|---|
| Intentional | 16.9 | 27.0 |
| Playing with Heat Source | 0.0 | 0.0 |
| Smoking | 6.8 | 10.8 |
| Heating | 0.0 | 0.0 |
| Cooking | 0.0 | 0.0 |
| Electrical Malfunction | 1.7 | 2.7 |
| Appliances | 1.7 | 2.7 |
| Open Flame | 3.4 | 5.4 |
| Other Heat | 3.4 | 5.4 |
| Other Equipment | 8.5 | 13.5 |
| Natural | 1.7 | 2.7 |
| Exposure | 1.7 | 2.7 |
| Equipment Misoperation, Failure | 1.7 | 2.7 |
| Other Unintentional, Careless | 11.9 | 18.9 |
| Investigation with Arson Module | 3.4 | 5.4 |
| Unknown | 37.3 | 0.0 |

Note: The distribution of causes should be viewed with caution due to the small number of fatal fires and the large percentage of fatal fires with unknown cause.

continued on next page

# Figure 22. Causes of Nonresidential Structure Fires and Fire Losses (2007)—Continued

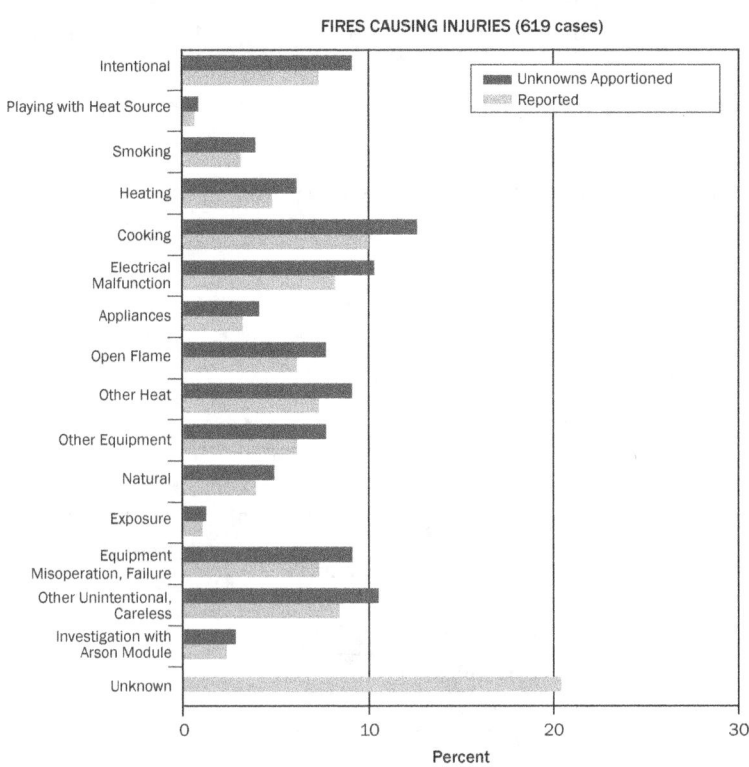

| Cause | Reported | Unknowns Apportioned |
|---|---|---|
| Intentional | 7.3 | 9.1 |
| Playing with Heat Source | 0.6 | 0.8 |
| Smoking | 3.1 | 3.9 |
| Heating | 4.8 | 6.1 |
| Cooking | 10.0 | 12.6 |
| Electrical Malfunction | 8.2 | 10.3 |
| Appliances | 3.2 | 4.1 |
| Open Flame | 6.1 | 7.7 |
| Other Heat | 7.3 | 9.1 |
| Other Equipment | 6.1 | 7.7 |
| Natural | 3.9 | 4.9 |
| Exposure | 1.0 | 1.2 |
| Equipment Misoperation, Failure | 7.3 | 9.1 |
| Other Unintentional, Careless | 8.4 | 10.5 |
| Investigation with Arson Module | 2.3 | 2.8 |
| Unknown | 20.4 | 0.0 |

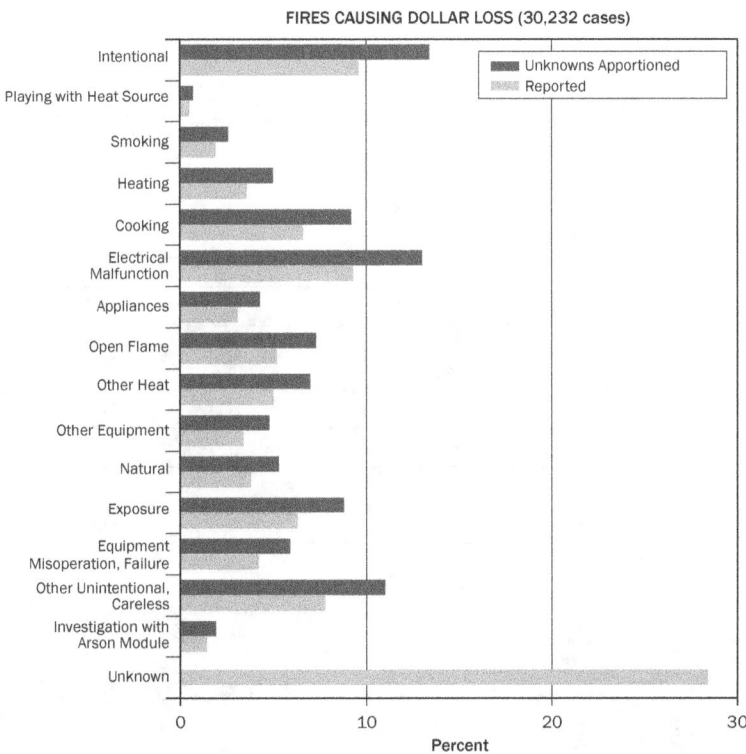

| Cause | Reported | Unknowns Apportioned |
|---|---|---|
| Intentional | 9.6 | 13.4 |
| Playing with Heat Source | 0.5 | 0.7 |
| Smoking | 1.9 | 2.6 |
| Heating | 3.6 | 5.0 |
| Cooking | 6.6 | 9.2 |
| Electrical Malfunction | 9.3 | 13.0 |
| Appliances | 3.1 | 4.3 |
| Open Flame | 5.2 | 7.3 |
| Other Heat | 5.0 | 7.0 |
| Other Equipment | 3.4 | 4.8 |
| Natural | 3.8 | 5.3 |
| Exposure | 6.3 | 8.8 |
| Equipment Misoperation, Failure | 4.2 | 5.9 |
| Other Unintentional, Careless | 7.8 | 11.0 |
| Investigation with Arson Module | 1.4 | 1.9 |
| Unknown | 28.4 | 0.0 |

Source: NFIRS.

## *Vehicle*

As shown in Figure 23, unintentionally set fires are the leading cause of fires, fatal fires, fires causing injuries, and fires causing dollar loss in vehicles (32, 48, 51, and 30 percent, respectively). Failure of equipment or heat source is the second leading cause in all categories except fatal fires (fires—24 percent, fires causing injuries—19 percent, fires causing dollar loss—27 percent) where the second leading cause is cause under investigation (28 percent).

### Figure 23. Causes of Vehicle Fires and Fire Losses (2007)

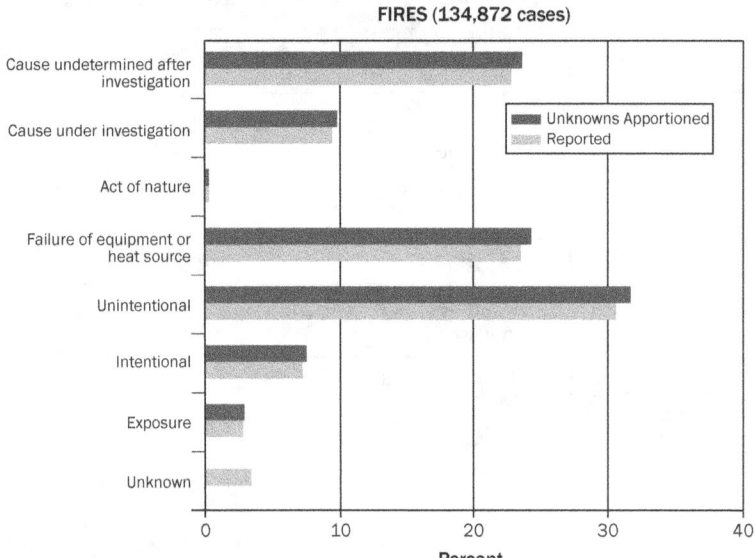

| Cause | Reported | Unknowns Apportioned |
|---|---|---|
| Cause undetermined after investigation | 22.8 | 23.6 |
| Cause under investigation | 9.4 | 9.8 |
| Act of nature | 0.3 | 0.3 |
| Failure of equipment or heat source | 23.5 | 24.3 |
| Unintentional | 30.6 | 31.7 |
| Intentional | 7.2 | 7.5 |
| Exposure | 2.8 | 2.9 |
| Unknown | 3.4 | 0.0 |

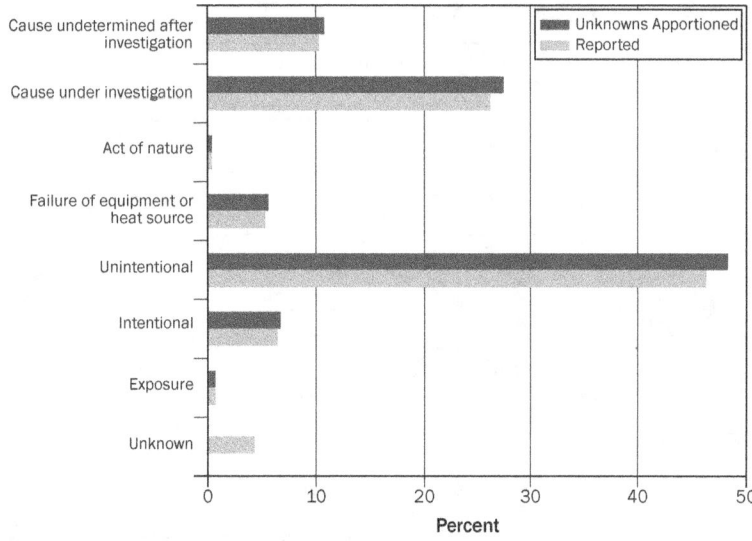

| Cause | Reported | Unknowns Apportioned |
|---|---|---|
| Cause undetermined after investigation | 10.3 | 10.8 |
| Cause under investigation | 26.3 | 27.5 |
| Act of nature | 0.4 | 0.4 |
| Failure of equipment or heat source | 5.3 | 5.6 |
| Unintentional | 46.3 | 48.3 |
| Intentional | 6.4 | 6.7 |
| Exposure | 0.7 | 0.7 |
| Unknown | 4.3 | 0.0 |

*continued on next page*

## Figure 23. Causes of Vehicle Fires and Fire Losses (2007)—Continued

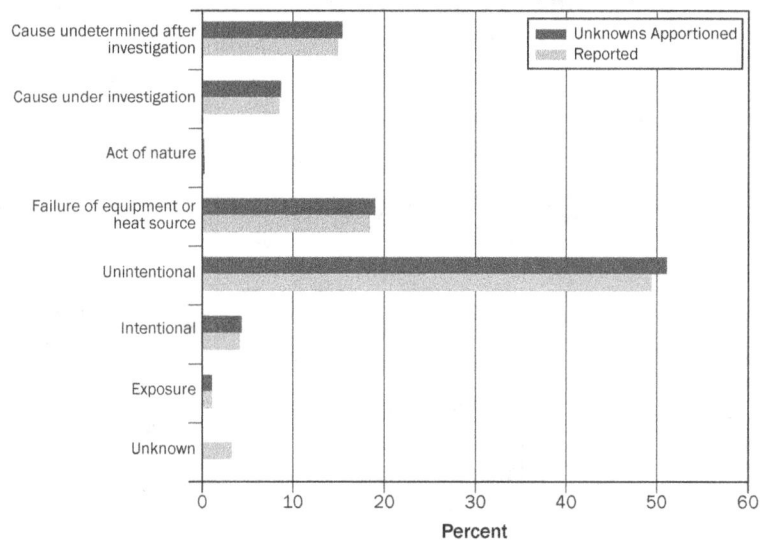

**FIRES CAUSING INJURIES (544 cases)**

| Cause | Reported | Unknowns Apportioned |
|---|---|---|
| Cause undetermined after investigation | 14.9 | 15.4 |
| Cause under investigation | 8.5 | 8.7 |
| Act of nature | 0.2 | 0.2 |
| Failure of equipment or heat source | 18.4 | 19.0 |
| Unintentional | 49.4 | 51.1 |
| Intentional | 4.2 | 4.4 |
| Exposure | 1.1 | 1.1 |
| Unknown | 3.3 | 0.0 |

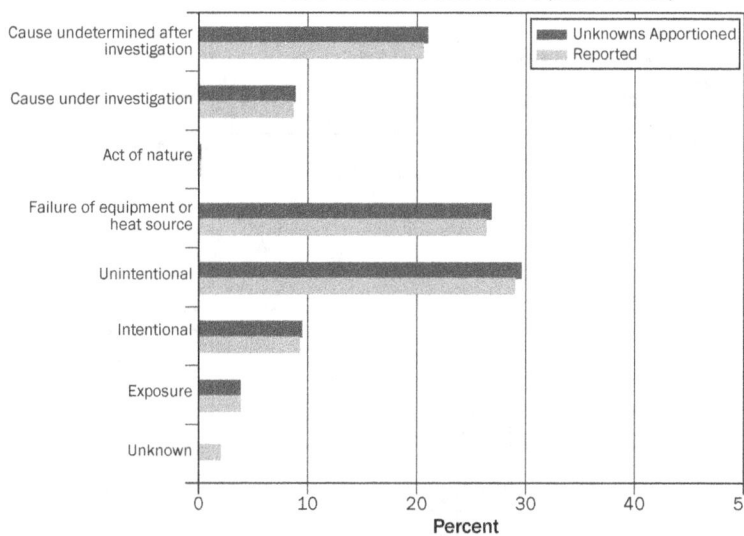

**FIRES CAUSING DOLLAR LOSS (72,577 cases)**

| Cause | Reported | Unknowns Apportioned |
|---|---|---|
| Cause undetermined after investigation | 20.6 | 21.0 |
| Cause under investigation | 8.7 | 8.9 |
| Act of nature | 0.2 | 0.2 |
| Failure of equipment or heat source | 26.3 | 26.8 |
| Unintentional | 29.0 | 29.6 |
| Intentional | 9.3 | 9.5 |
| Exposure | 3.9 | 3.9 |
| Unknown | 2.1 | 0.0 |

Source: NFIRS.

## *Outside*

Intentionally set and unintentionally set fires are the leading causes of fatal outside fires, both at 37 percent (Figure 24). Unintentionally set fires are also the leading cause of fires, fires causing injuries, and fires causing dollar loss in outside fires (38, 53, and 38 percent, respectively). Fire causes are reported as undetermined after an investigation for 31 percent of outside fires and 24 percent of outside fires causing dollar loss. Intentionally set fires are the second leading cause of outside fires causing injuries (23 percent).

### Figure 24. Causes of Outside Fires and Fire Losses (2007)

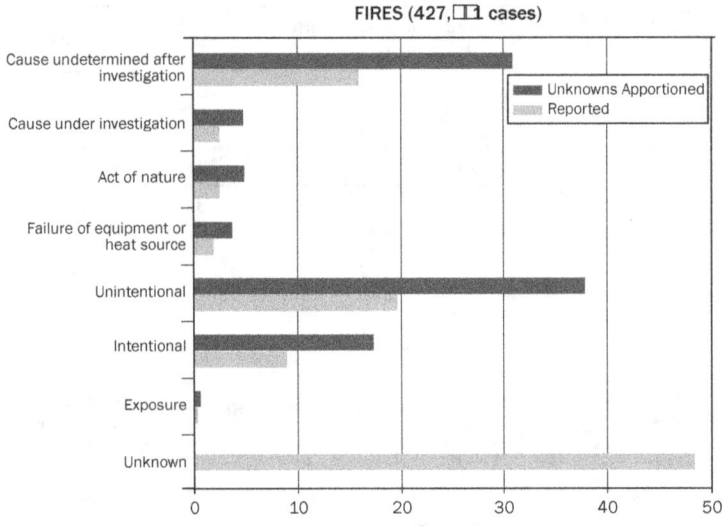

| Cause | Reported | Unknowns Apportioned |
|---|---|---|
| Cause undetermined after investigation | 15.9 | 30.9 |
| Cause under investigation | 2.5 | 4.8 |
| Act of nature | 2.5 | 4.9 |
| Failure of equipment or heat source | 1.9 | 3.7 |
| Unintentional | 19.6 | 37.9 |
| Intentional | 8.9 | 17.3 |
| Exposure | 0.3 | 0.6 |
| Unknown | 48.4 | 0.0 |

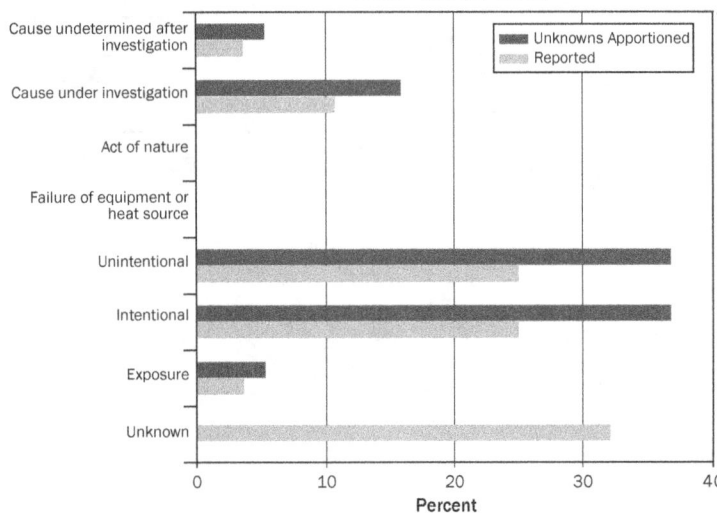

| Cause | Reported | Unknowns Apportioned |
|---|---|---|
| Cause undetermined after investigation | 3.6 | 5.3 |
| Cause under investigation | 10.7 | 15.8 |
| Act of nature | 0.0 | 0.0 |
| Failure of equipment or heat source | 0.0 | 0.0 |
| Unintentional | 25.0 | 36.8 |
| Intentional | 25.0 | 36.8 |
| Exposure | 3.6 | 5.3 |
| Unknown | 32.1 | 0.0 |

Note: The distribution of causes should be viewed with caution due to the small number of fatal fires and the large percentage of fatal fires with unknown cause.

*continued on next page*

# Figure 24. Causes of Outside Fires and Fire Losses (2007)—Continued

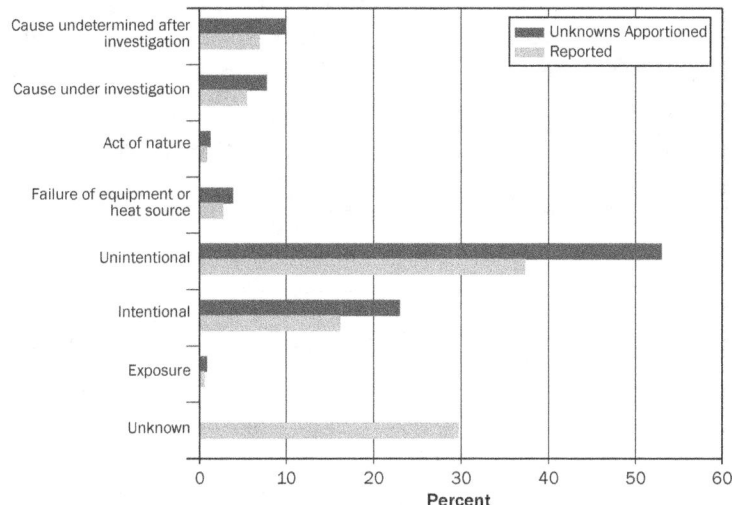

FIRES CAUSING INJURIES (327 cases)

| Cause | Reported | Unknowns Apportioned |
|---|---|---|
| Cause undetermined after investigation | 7.0 | 10.0 |
| Cause under investigation | 5.5 | 7.8 |
| Act of nature | 0.9 | 1.3 |
| Failure of equipment or heat source | 2.8 | 3.9 |
| Unintentional | 37.3 | 53.0 |
| Intentional | 16.2 | 23.0 |
| Exposure | 0.6 | 0.9 |
| Unknown | 29.7 | 0.0 |

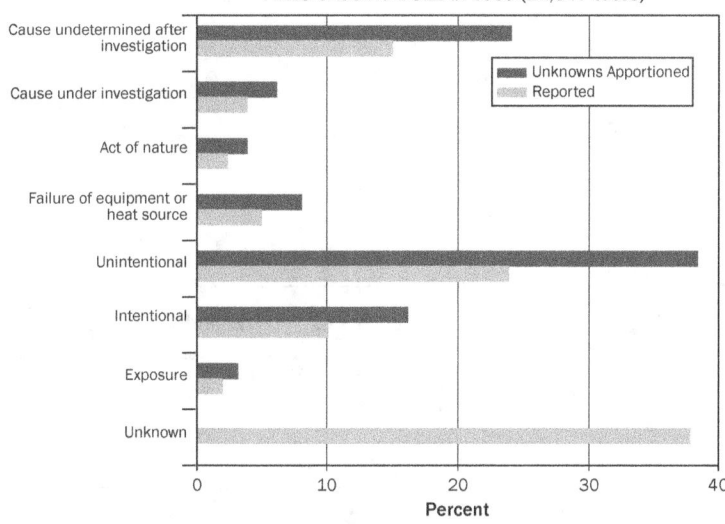

FIRES CAUSING DOLLAR LOSS (2☐847 cases)

| Cause | Reported | Unknowns Apportioned |
|---|---|---|
| Cause undetermined after investigation | 15.0 | 24.1 |
| Cause under investigation | 3.9 | 6.2 |
| Act of nature | 2.4 | 3.9 |
| Failure of equipment or heat source | 5.0 | 8.1 |
| Unintentional | 23.9 | 38.4 |
| Intentional | 10.1 | 16.2 |
| Exposure | 2.0 | 3.2 |
| Unknown | 37.8 | 0.0 |

Source:  NFIRS.

## *Other*

Just as with vehicle and outside fires, unintentionally set fires are the leading cause of other fires, fatal fires, fires causing injuries, and fires causing dollar loss (41, 40, 66, and 40 percent, respectively) (Figure 25). Failure of equipment or heat source is the second leading cause of other fires (27 percent), other fires causing injuries (11 percent), and other fires causing dollar loss (31 percent). In 24 percent of fatal fires, the cause of the fire is under investigation at the time the fire incident report was submitted. For an additional 24 percent of fatal fires, the cause is undetermined after the investigation.

### Figure 25. Causes of Other Fires and Fire Losses (2007)

FIRES (77,2□3 cases)

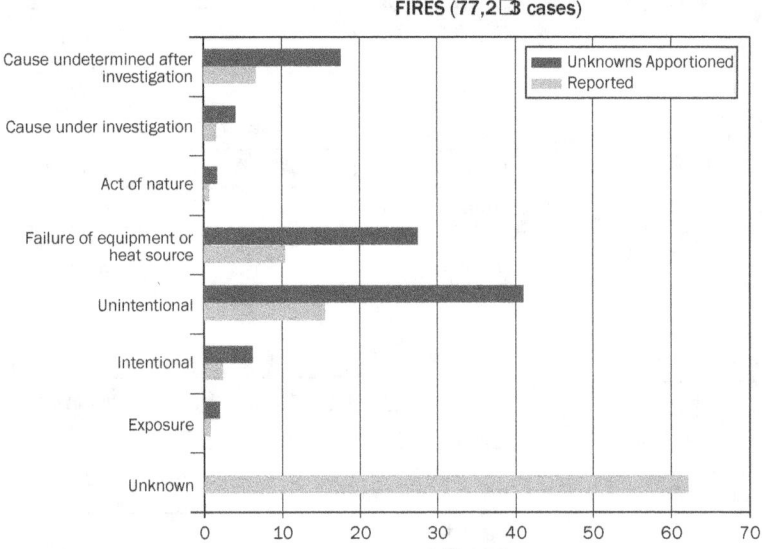

| Cause | Reported | Unknowns Apportioned |
|---|---|---|
| Cause undetermined after investigation | 6.7 | 17.6 |
| Cause under investigation | 1.6 | 4.1 |
| Act of nature | 0.7 | 1.7 |
| Failure of equipment or heat source | 10.4 | 27.4 |
| Unintentional | 15.5 | 41.0 |
| Intentional | 2.3 | 6.1 |
| Exposure | 0.8 | 2.0 |
| Unknown | 62.1 | 0.0 |

FATAL FIRES (38 cases)

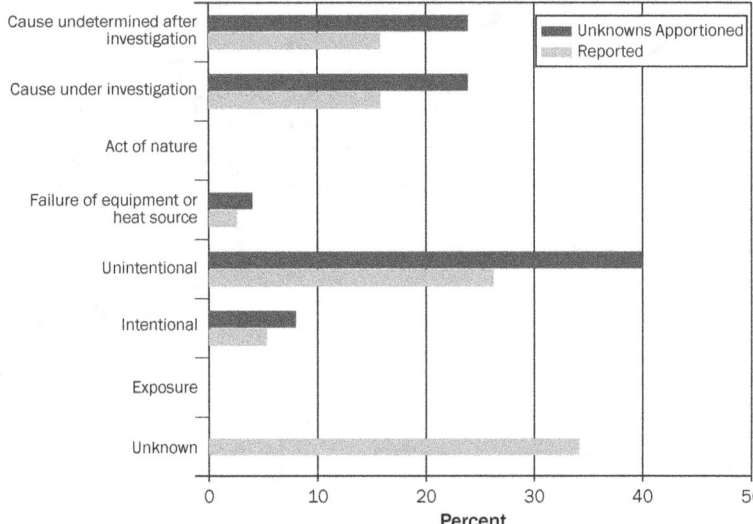

| Cause | Reported | Unknowns Apportioned |
|---|---|---|
| Cause undetermined after investigation | 15.8 | 24.0 |
| Cause under investigation | 15.8 | 24.0 |
| Act of nature | 0.0 | 0.0 |
| Failure of equipment or heat source | 2.6 | 4.0 |
| Unintentional | 26.3 | 40.0 |
| Intentional | 5.3 | 8.0 |
| Exposure | 0.0 | 0.0 |
| Unknown | 34.2 | 0.0 |

Note:   The distribution of causes should be viewed with caution due to the small number of fatal fires and the large percentage of fatal fires with unknown cause.

*continued on next page*

# Figure 25. Causes of Other Fires and Fire Losses (2007)—Continued

**FIRES CAUSING INJURIES (3□2 cases)**

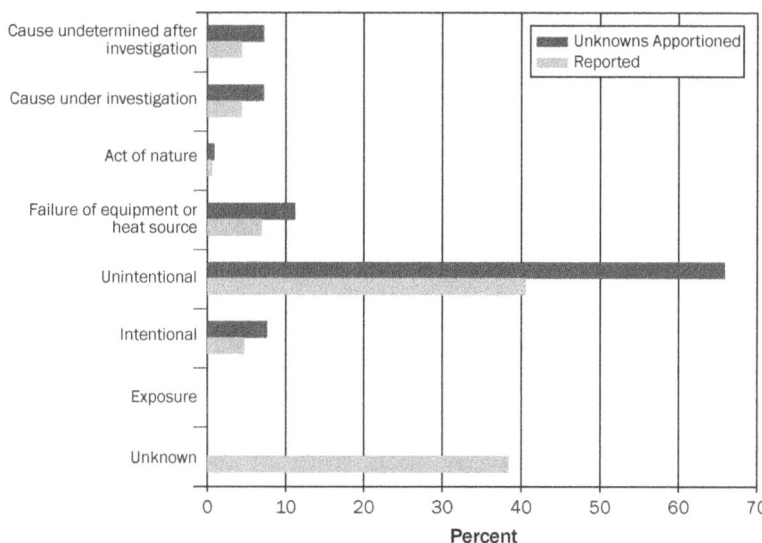

| Cause | Reported | Unknowns Apportioned |
|---|---|---|
| Cause undetermined after investigation | 4.4 | 7.2 |
| Cause under investigation | 4.4 | 7.2 |
| Act of nature | 0.6 | 0.9 |
| Failure of equipment or heat source | 6.9 | 11.2 |
| Unintentional | 40.6 | 65.9 |
| Intentional | 4.7 | 7.6 |
| Exposure | 0.0 | 0.0 |
| Unknown | 38.4 | 0.0 |

**FIRES CAUSING DOLLAR LOSS (14,81□cases)**

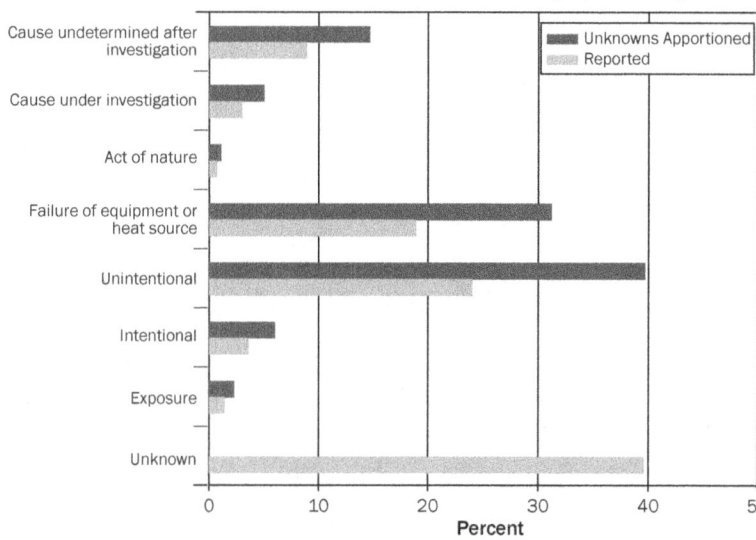

| Cause | Reported | Unknowns Apportioned |
|---|---|---|
| Cause undetermined after investigation | 8.9 | 14.7 |
| Cause under investigation | 3.0 | 5.0 |
| Act of nature | 0.7 | 1.1 |
| Failure of equipment or heat source | 18.9 | 31.2 |
| Unintentional | 24.0 | 39.7 |
| Intentional | 3.6 | 6.0 |
| Exposure | 1.4 | 2.3 |
| Unknown | 39.6 | 0.0 |

Source: NFIRS.

# Chapter 3

## Buildings and Other Properties

This chapter provides an overview of the fire problem in buildings, vehicle and other mobile properties, and outside and other properties over the 5-year period from 2003 to 2007, with specific focus on 2007. Stand-alone reports for residential and nonresidential buildings have been published with more detail.

## BUILDINGS

This analysis of building fires is discussed in two major sections: residential (including one- and two-family dwellings, multifamily dwellings, and other residential buildings) and nonresidential (including industrial and commercial properties, institutions, educational establishments, mobile properties, and storage properties).

### *Residential*

The residential building category is discussed in four subsections: an overview of all residential buildings, one- and two-family dwellings, multifamily dwellings, and other residential buildings and property uses reported as residences.

### Overview of All Residential Buildings

The residential building portion of the fire problem continues to account for the vast majority of civilian casualties. National estimates for 2007 show that 81 percent of fire deaths and 77 percent of fire injuries occur in residential buildings, and account for over half (56 percent) of the total dollar loss from all fires. These losses result from 25 percent of reported fires.[1,2]

The term "residential buildings" includes what are commonly referred to as "homes," whether they are one- or two-family dwellings or multifamily buildings. It also includes manufactured housing, hotels and motels, residential hotels, dormitories, assisted living facilities, and halfway houses—residences

---

[1] National estimates for building fires are derived from data from NFIRS and the NFPA's annual fire loss survey (Karter, Jr., Michael J., *Fire Loss in the United States 2007*, NFPA, August 2008). The percentages shown here are derived from the national estimates of residential building fires and the summary data presented in the NFPA report.

[2] Karter, Jr., Michael J., *Fire Loss in the United States 2007*, NFPA, August 2008. These percentages are derived from summary data presented in this report. Dollar loss percentages do not include the losses from the 2007 California Fire Storm as those losses were not ascribed to property types.

for formerly institutionalized individuals (mentally impaired patients, drug addicts, or convicts) that are designed to facilitate their readjustment to private life. The term "residential buildings" does not include institutions such as prisons, nursing homes, juvenile care facilities, or hospitals, though many people may reside in them for short or long periods of time.

Figure 26, based on national estimates of the residential building fire problem, shows the 5-year trend in residential fires, deaths, injuries, and dollar loss. The trend in number of residential building fires increased 2 percent, while residential building fire deaths and injuries declined 14, and 3 percent, respectively. The decreases in fire deaths and fire injuries continue the downward trends estimated in past editions of this report. However, while the estimated number of residential building fires increased 2 percent in the past 5 years, the number of occupied residences has increased 5 percent. The 2 percent increase in residential building fires should therefore be considered in the context of the increase in occupied housing, as it is likely that fires have actually declined per household.[3] Dollar losses (adjusted for inflation), increased 14 percent over the 2003 to 2007 period.

Because residential building fires resulted in an average of 2,843 civilian deaths, 13,305 injuries, and $6.7 billion in dollar loss (adjusted to 2007 dollars) over the 5-year period, the fire problem in U.S. residences is of significant concern.

## One- and Two-Family Dwellings

One- and two-family dwellings are where 73 percent of the people in the United States reside.[4] The residential building fire profile is, therefore, dominated by this category. Manufactured housing (mobile homes used as fixed residences) is included here in the profile for one- and two-family homes.

### Trends

As with the residential trends, one- and two-family dwelling fires increased (4 percent), while deaths and injuries declined during the 5-year period (15 and 5 percent respectively). Dollar loss increased (16 percent), as shown in Figure 27. The increased use of smoke alarms are credited as a major factor in the reduction in the number of reported fires; without smoke alarms, the increase in one- and two-family dwelling fires may have been larger.

---

[3] In 2003, there were an estimated 105,842,000 occupied residences (http://www.census.gov/hhes/www/housing/ahs/ahs03/tab21.htm). In 2007, there were an estimated 110,692,000 occupied residences (http://www.census.gov/hhes/www/housing/ahs/ahs07/tab2-1.xls).

[4] The U.S. Census Bureau shows that, in 2007, 76.3 percent (84.4 million) of occupied housing units were one-unit attached and detached structures or mobile homes (http://www.census.gov/hhes/www/housing/ahs/ahs07/tab1a-1.xls for occupied housing). Household size was estimated at 2.6 people per household (http://factfinder.census.gov/servlet/ACSSAFFFacts?_submenuId=factsheet_1&_sse=on). Thus, 84.4 million housing units x 2.6 people per household = 219.4 million. With the 2007 U.S. population given as 301.3 million, (http://www.census.gov/popest/national/asrh/NC-EST2008/NC-EST2008-03.xls), approximately 72.8 percent of the population lived in what NFIRS defines as one- and two-family housing.

## Figure 26. Trends in Residential Building Fires and Fire Losses

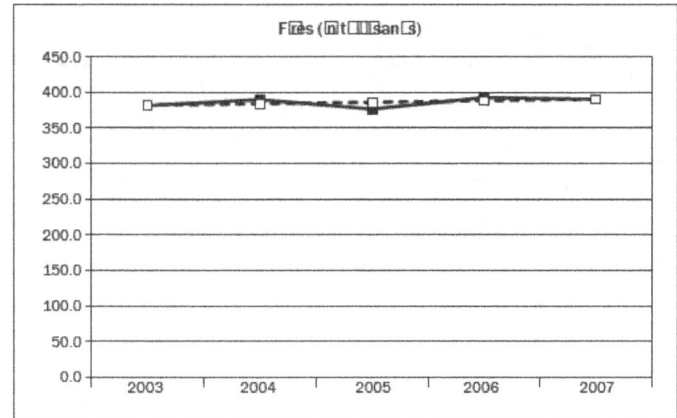

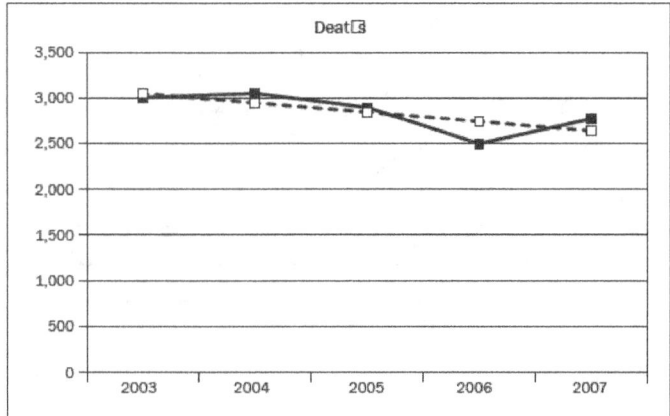

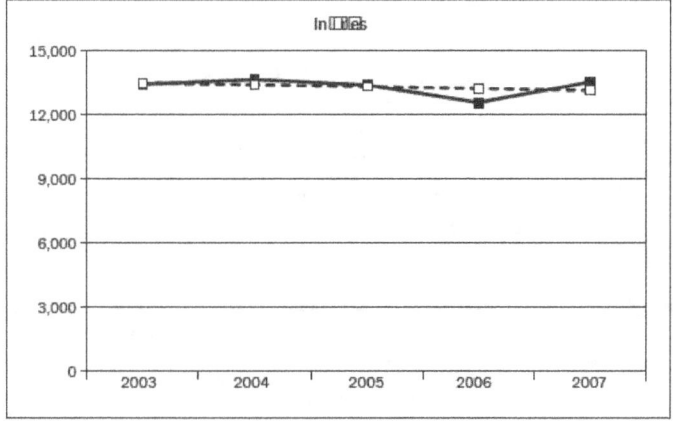

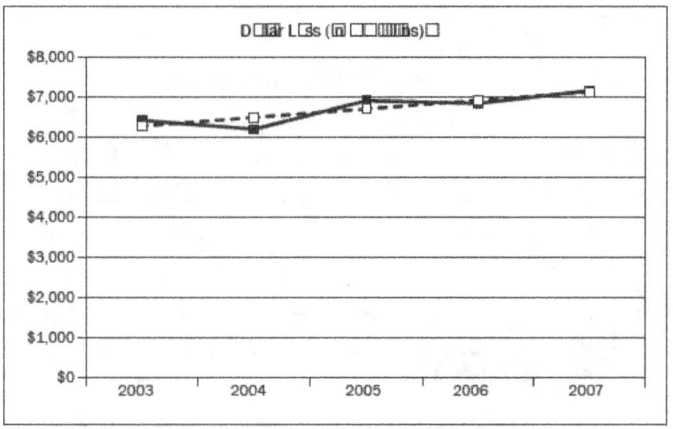

| FIRES (THOUSANDS) | |
| --- | --- |
| Year | Value |
| 2003 | 381.3 |
| 2004 | 389.8 |
| 2005 | 376.5 |
| 2006 | 392.7 |
| 2007 | 390.3 |
| 5-Year Trend (%) | 2.2% |

| DEATHS | |
| --- | --- |
| Year | Value |
| 2003 | 3,005 |
| 2004 | 3,050 |
| 2005 | 2,895 |
| 2006 | 2,495 |
| 2007 | 2,770 |
| 5-Year Trend (%) | -13.5% |

| INJURIES | |
| --- | --- |
| Year | Value |
| 2003 | 13,425 |
| 2004 | 13,650 |
| 2005 | 13,375 |
| 2006 | 12,550 |
| 2007 | 13,525 |
| 5-Year Trend (%) | -2.7% |

| DOLLAR LOSS ($M)* *ADJUSTED TO 2007 DOLLARS | |
| --- | --- |
| Year | Value |
| 2003 | $6,413 |
| 2004 | $6,196 |
| 2005 | $6,906 |
| 2006 | $6,835 |
| 2007 | $7,157 |
| 5-Year Trend (%) | 13.6% |

Sources: NFPA, NFIRS, and Consumer Price Index.

## Figure 27. Trends in One- and Two-Family Dwelling Fires and Fire Losses

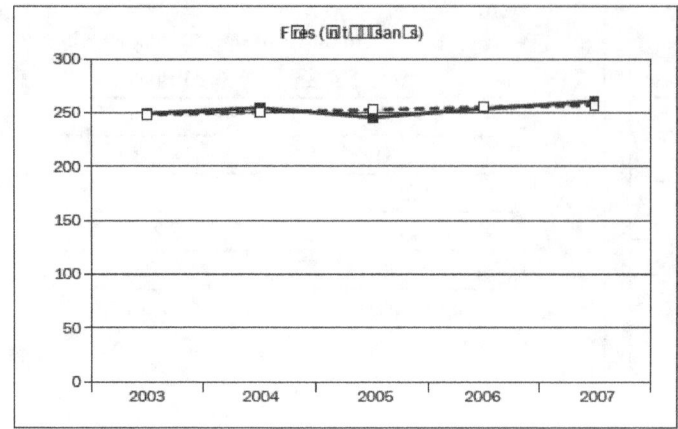

| FIRES (THOUSANDS) | |
| --- | --- |
| Year | Value |
| 2003 | 249.4 |
| 2004 | 254.6 |
| 2005 | 245.9 |
| 2006 | 253.8 |
| 2007 | 260.8 |
| **5-Year Trend (%)** | **3.5%** |

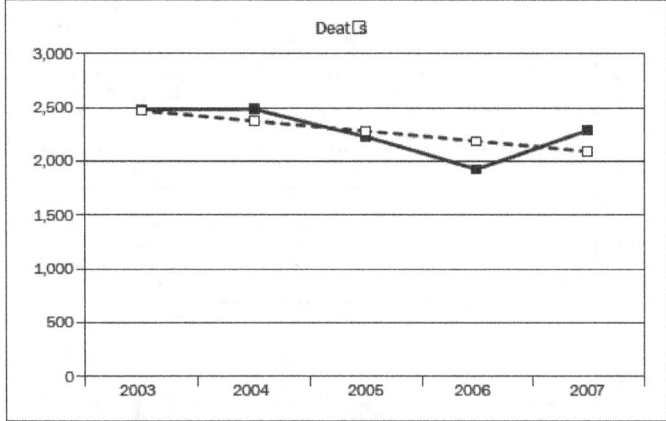

| DEATHS | |
| --- | --- |
| Year | Value |
| 2003 | 2,480 |
| 2004 | 2,485 |
| 2005 | 2,225 |
| 2006 | 1,925 |
| 2007 | 2,285 |
| **5-Year Trend (%)** | **-15.4%** |

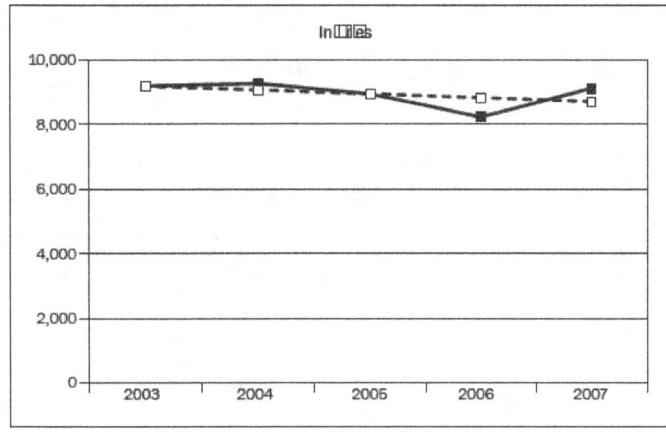

| INJURIES | |
| --- | --- |
| Year | Value |
| 2003 | 9,200 |
| 2004 | 9,275 |
| 2005 | 8,950 |
| 2006 | 8,225 |
| 2007 | 9,125 |
| **5-Year Trend (%)** | **-5.2%** |

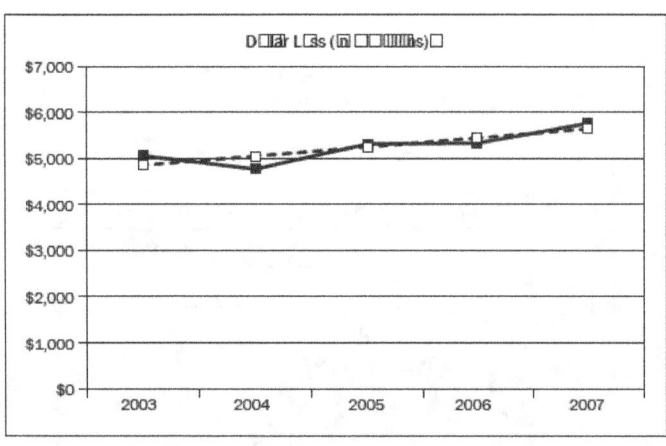

| DOLLAR LOSS ($M)* *ADJUSTED TO 2007 DOLLARS | |
| --- | --- |
| Year | Value |
| 2003 | $5,055 |
| 2004 | $4,766 |
| 2005 | $5,310 |
| 2006 | $5,327 |
| 2007 | $5,755 |
| **5-Year Trend (%)** | **16.2%** |

Sources: NFPA, NFIRS, and Consumer Price Index.

## *Multifamily Dwellings*

Multifamily dwellings, formerly referred to as "apartments" in previous editions of this report, tend to be regulated by stricter building codes than one- and two-family dwellings. Multifamily dwellings include condominiums, townhouses, rowhouses, and tenements, as well as traditional apartments (lowrise or highrise apartments). Many multifamily dwellings are rental properties, often falling under more stringent fire prevention statutes. Many multifamily-dwelling communities also have a different socioeconomic mix of residents compared to single-family-dwelling communities. They may have more low-income families in housing projects or more high-income families in luxury highrises, or they may be centers of living for the elderly. In large cities, all of these groups are represented in multifamily dwellings.

### Trends

Figure 28 shows the 5-year trends in multifamily dwelling fires and losses.[5] The number of multifamily dwelling fires declined modestly (2 percent), though the number of deaths and injuries in multifamily dwelling fires increased 7 and 4 percent, respectively. These trends appear to be in opposition to the one- and two-family residence trends. Though one- and two-family residences experienced a 4 percent increase in the total number of fires, they benefitted from a notable decrease in both the number of deaths and injuries (15 and 5 percent, respectively). Conversely, while multifamily dwellings may have experienced a minor decrease in total fires, they unfortunately experienced an increase in deaths and injuries. Dollar losses in both types of buildings continued to increase: adjusted dollar losses were up 6 percent in multifamily dwellings and 16 percent in one- and two-family dwellings during the 5-year period.

The increase in multifamily dwelling deaths and injuries is surprising given that multifamily buildings tend to have stricter building codes including the presence of smoke alarms and sprinkler systems. More detailed study of socioeconomic and demographic changes over time may reveal some of the other factors involved in fire incidence.

## *Other Residential Buildings*

Other residential buildings include rooming houses, dormitories, residential hotels, halfway houses, hotels and motels, and miscellaneous and unclassified buildings reported as residences. This category does not include homes for the elderly, prisons, or other institutions. These categories are addressed as part of nonresidential buildings in the next section.

### Trends

Figure 29 shows an increase (9 percent) in the number of other residential fires, while showing a substantial decrease in the number of fire deaths (35 percent). Injuries decreased to a lesser degree, down 7 percent. Fire deaths ranged from 80 to 160 per year; injuries ranged from 525 to 600.[6] Adjusted dollar loss has decreased less than 1 percent over 5 years.

---

[5] These national estimates of multifamily residential building fires and losses vary substantially from the NFPA estimates. The estimates shown reflect the proportion of multifamily residential building fires and losses as collected in the NFIRS data set. This topic is addressed further in Appendix A.

[6] These national estimates of other residential building fires and losses vary substantially from the NFPA estimates. The estimates shown reflect the proportion of other residential building fires and losses as collected in the NFIRS data set. This topic is addressed further in Appendix A.

## Figure 28. Trends in Multifamily Dwelling Fires and Fire Losses

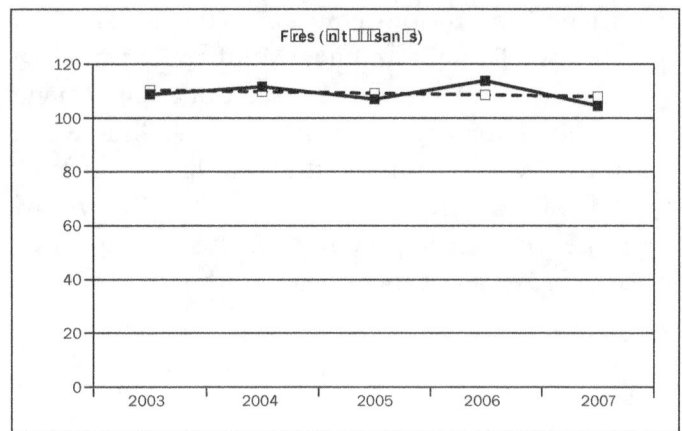

| FIRES (THOUSANDS) | |
|---|---|
| Year | Value |
| 2003 | 108.8 |
| 2004 | 111.7 |
| 2005 | 107.0 |
| 2006 | 113.9 |
| 2007 | 104.6 |
| 5-Year Trend (%) | -2.2% |

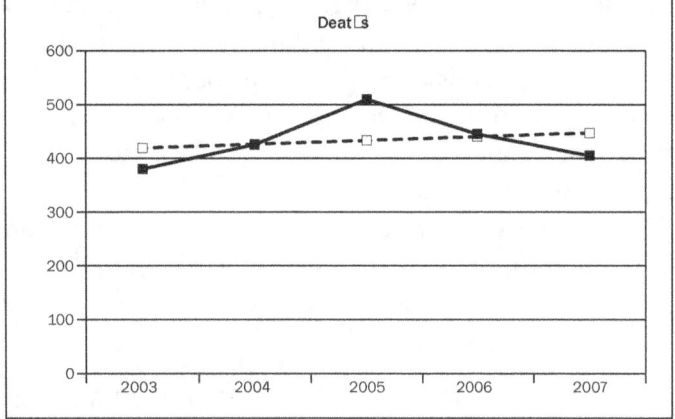

| DEATHS | |
|---|---|
| Year | Value |
| 2003 | 380 |
| 2004 | 425 |
| 2005 | 510 |
| 2006 | 445 |
| 2007 | 405 |
| 5-Year Trend (%) | 6.7% |

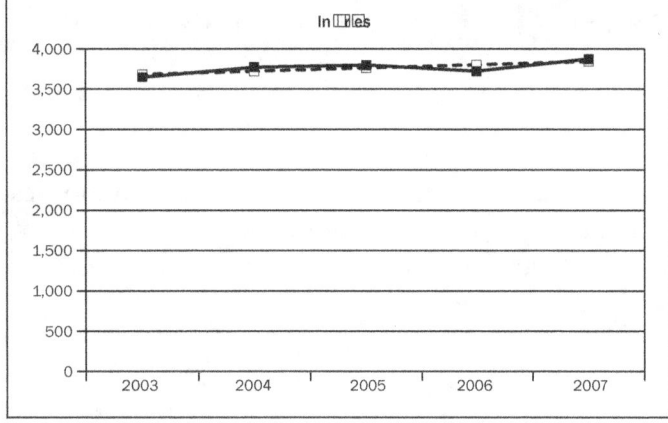

| INJURIES | |
|---|---|
| Year | Value |
| 2003 | 3,650 |
| 2004 | 3,775 |
| 2005 | 3,800 |
| 2006 | 3,725 |
| 2007 | 3,875 |
| 5-Year Trend (%) | 4.3% |

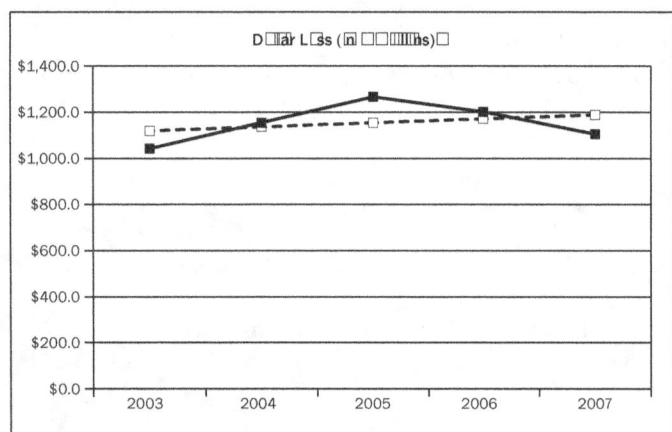

| DOLLAR LOSS ($M)* | |
|---|---|
| *ADJUSTED TO 2007 DOLLARS | |
| Year | Value |
| 2003 | $1,043 |
| 2004 | $1,155 |
| 2005 | $1,267 |
| 2006 | $1,202 |
| 2007 | $1,106 |
| 5-Year Trend (%) | 6.2% |

Sources:  NFPA, NFIRS, and Consumer Price Index.

## Figure 29. Trends in Other Residential Building Fires and Fire Losses

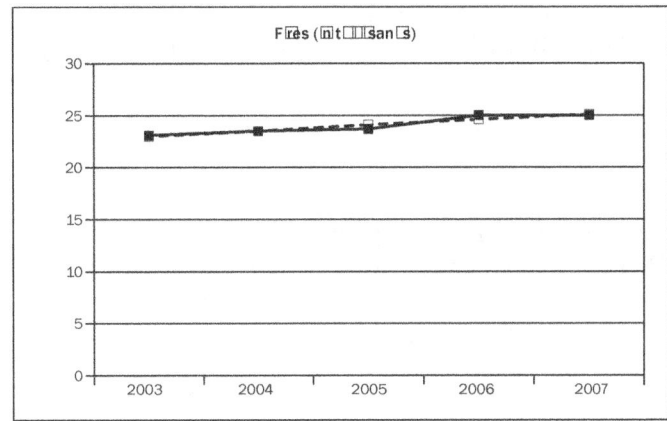

**FIRES (THOUSANDS)**

| Year | Value |
| --- | --- |
| 2003 | 23.1 |
| 2004 | 23.5 |
| 2005 | 23.7 |
| 2006 | 25.0 |
| 2007 | 25.0 |
| **5-Year Trend (%)** | **9.2%** |

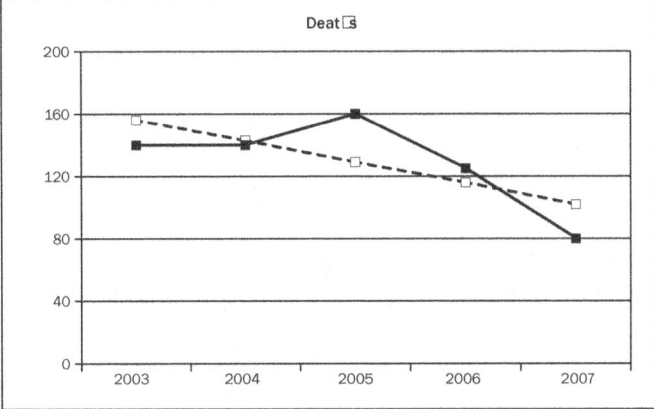

**DEATHS**

| Year | Value |
| --- | --- |
| 2003 | 140 |
| 2004 | 140 |
| 2005 | 160 |
| 2006 | 125 |
| 2007 | 80 |
| **5-Year Trend (%)** | **-34.6%** |

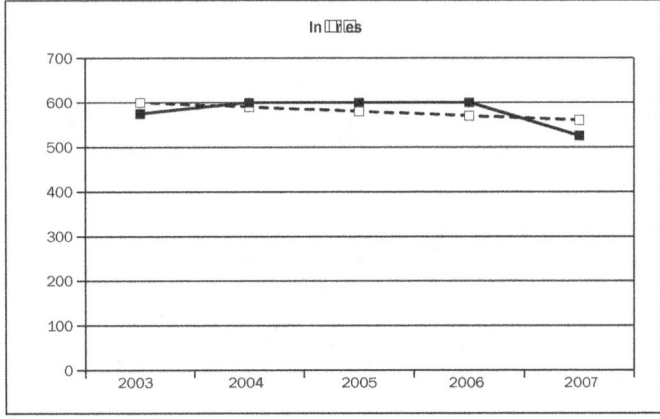

**INJURIES**

| Year | Value |
| --- | --- |
| 2003 | 575 |
| 2004 | 600 |
| 2005 | 600 |
| 2006 | 600 |
| 2007 | 525 |
| **5-Year Trend (%)** | **-6.7%** |

**DOLLAR LOSS ($M)***
*ADJUSTED TO 2007 DOLLARS

| Year | Value |
| --- | --- |
| 2003 | $314 |
| 2004 | $276 |
| 2005 | $329 |
| 2006 | $306 |
| 2007 | $296 |
| **5-Year Trend (%)** | **-0.8%** |

Sources: NFPA, NFIRS, and Consumer Price Index.

## *Nonresidential Buildings*

The nonresidential building category includes industrial and commercial properties, institutions (such as hospitals, nursing homes, and prisons), educational establishments (from preschool through university), mobile properties, and storage properties.

## Trends

Substantial public and private fire prevention efforts have focused on protecting nonresidential buildings. The results have proven effective in the main, especially relative to the residential fire problem. National estimates of nonresidential building fires and losses annually account for only 7 percent of fires, 3 percent of deaths, and 7 percent of injuries. These properties, however, account for a disproportionately large annual dollar loss, 22 percent.

The 5-year trends for fires, deaths, and injuries decreased during the 2003 to 2007 period.

Figure 30 shows the downward trend for each of these measures (fires, 2 percent; deaths, 57 percent; injuries, 3 percent). The trend for dollar loss, however, increased 12 percent over the 5-year period.

There were an estimated 90 deaths in nonresidential building fires in 2007. The 2003 peak (185 deaths) includes 100 deaths in The Station nightclub fire in Rhode Island, and 31 deaths in nursing home fires in Connecticut and Tennessee. The dollar loss peak in 2007 may be due to an increase in nonresidential large-loss fires. The NFPA estimates the number of large-loss fires in 2007 reached a total of 71, an additional 26 large-loss fires than in 2006, 36 of which were nonresidential building fires.[7]

---

[7] Badger, Stephen G., *Large Loss for 2007*, NFPA, November/December 2008 and Badger, Stephen G., *Large Loss for 2006*, NFPA, November/December 2007.

# Figure 30. Trends in Nonresidential Building Fires and Fire Losses

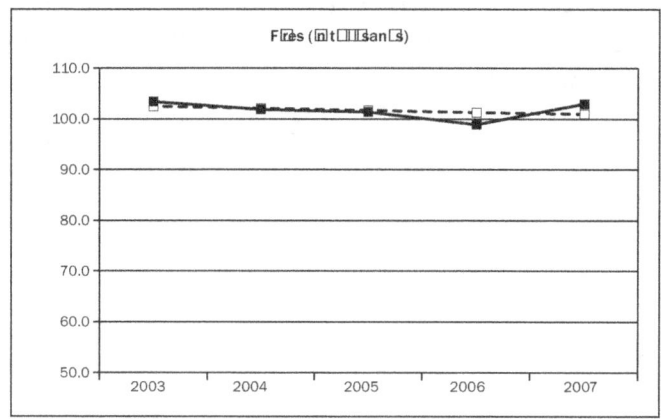

**FIRES (THOUSANDS)**

| Year | Value |
|---|---|
| 2003 | 103.4 |
| 2004 | 101.9 |
| 2005 | 101.4 |
| 2006 | 98.9 |
| 2007 | 103.0 |
| **5-Year Trend (%)** | **-1.5%** |

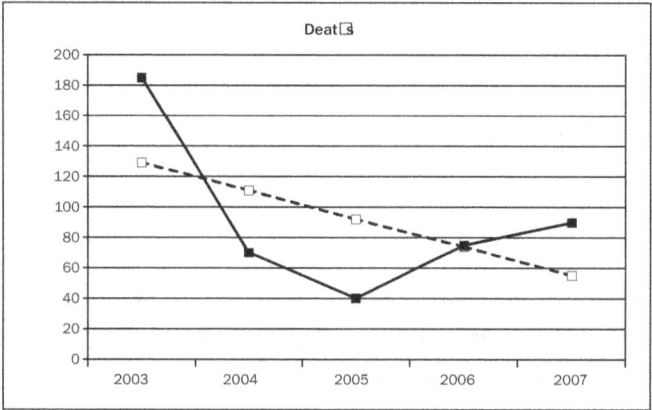

**DEATHS**

| Year | Value |
|---|---|
| 2003 | 185 |
| 2004 | 70 |
| 2005 | 40 |
| 2006 | 75 |
| 2007 | 90 |
| **5-Year Trend (%)** | **-57.4%** |

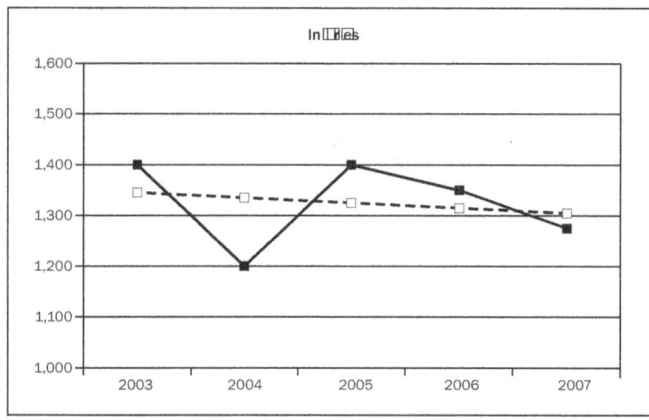

**INJURIES**

| Year | Value |
|---|---|
| 2003 | 1,400 |
| 2004 | 1,200 |
| 2005 | 1,400 |
| 2006 | 1,350 |
| 2007 | 1,275 |
| **5-Year Trend (%)** | **-3.0%** |

**DOLLAR LOSS ($M)***
*ADJUSTED TO 2007 DOLLARS

| Year | Value |
|---|---|
| 2003 | $2,566 |
| 2004 | $2,341 |
| 2005 | $2,270 |
| 2006 | $2,412 |
| 2007 | $2,868 |
| **5-Year Trend (%)** | **11.5%** |

Sources: NFPA, NFIRS, and Consumer Price Index.

# VEHICLES AND OTHER MOBILE PROPERTIES

Vehicle fires account for a larger portion of the fire problem than many people realize. In 2007, vehicles accounted for 11 percent of fire deaths overall, 9 percent of fire injuries, 11 percent of dollar losses, and 17 percent of all reported fires—nearly one in every six fires.[8]

The vast majority of fires, casualties, and dollar losses from mobile property involve cars and trucks, with cars clearly dominating this group. Fire departments respond to about as many fires involving vehicles as they do to fires involving one- and two-family residences.

## Trends

The trends in mobile property fires, deaths, injuries, and dollar losses are shown in Figure 31. The numbers of fires, deaths, and injuries and the dollar loss continue to decrease (16, 18, 4, and 8 percent, respectively), according to the NFPA estimates.

Figure 32 shows that the vast majority of mobile property fires and losses are from highway vehicles. The complexity and ambiguity in counting losses associated with accidents are described below in "Special Data Problems." The 5-year trends in highway vehicle fires, deaths, and dollar loss show substantial decreases (19, 20, and 13 percent, respectively).

## Special Data Problems

When there are fatalities associated with a mobile property accident such as a collision between two cars, it is often difficult to determine whether the fatalities were the result of the mechanical forces or the fire that ensued. Because of the very large number of vehicle fatalities occurring in this country each year and the frequency of fires associated with these accidents, there can be a substantial error in estimating the total number of fire deaths if this issue is not carefully addressed. A fire fatality should be counted only if a person was trapped and killed by the fire, rather than killed on impact and subsequently exposed to the fire.

---

[8] Karter, Jr., Michael J., *Fire Loss in the United States* 2007, NFPA, August 2008. These percentages are derived from summary data presented in this report. Dollar loss percentages do not include the losses from the 2007 California Fire Storm as those losses were not ascribed to property types.

# Figure 31. Trends in Mobile Property Fires and Fire Losses

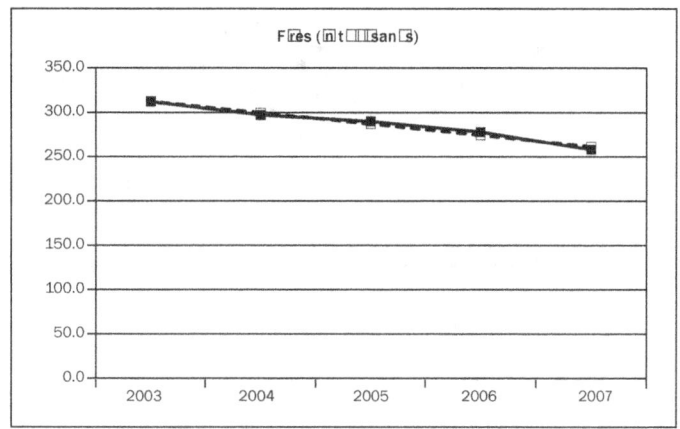

**FIRES (THOUSANDS)**

| Year | Value |
| --- | --- |
| 2003 | 312.0 |
| 2004 | 297.0 |
| 2005 | 290.0 |
| 2006 | 278.0 |
| 2007 | 258.0 |
| **5-Year Trend (%)** | **-16.3%** |

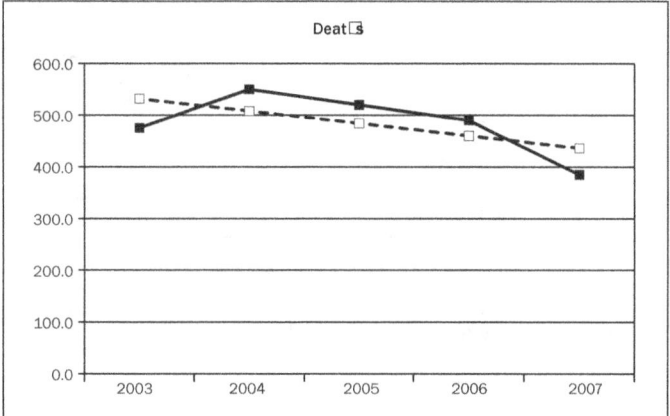

**DEATHS**

| Year | Value |
| --- | --- |
| 2003 | 475.0 |
| 2004 | 550.0 |
| 2005 | 520.0 |
| 2006 | 490.0 |
| 2007 | 385.0 |
| **5-Year Trend (%)** | **-18.0%** |

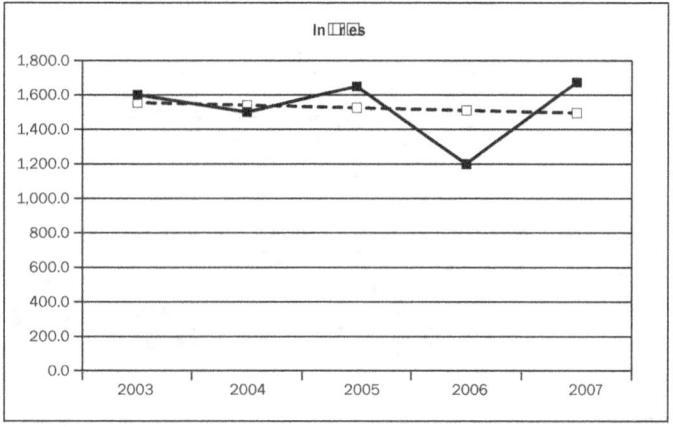

**INJURIES**

| Year | Value |
| --- | --- |
| 2003 | 1,600.0 |
| 2004 | 1,500.0 |
| 2005 | 1,650.0 |
| 2006 | 1,200.0 |
| 2007 | 1,675.0 |
| **5-Year Trend (%)** | **-3.9%** |

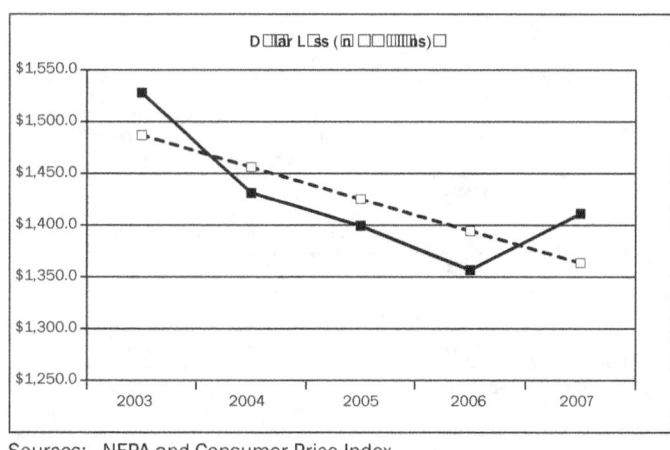

**DOLLAR LOSS ($M)***
*ADJUSTED TO 2007 DOLLARS

| Year | Value |
| --- | --- |
| 2003 | $1,528.0 |
| 2004 | $1,431.3 |
| 2005 | $1,399.3 |
| 2006 | $1,356.6 |
| 2007 | $1,411.0 |
| **5-Year Trend (%)** | **-8.3%** |

Sources: NFPA and Consumer Price Index.

## Figure 32. Trends in Highway Vehicle vs. Other Mobile Property Fires and Fire Losses

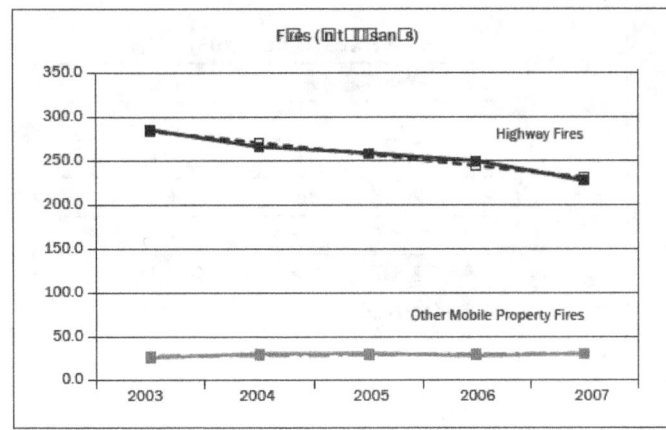

### FIRES (THOUSANDS)

| Year | Highway Vehicle Value | Other Mobile Value |
|------|------------------------|--------------------|
| 2003 | 286.0 | 26.0 |
| 2004 | 266.5 | 30.5 |
| 2005 | 259.0 | 31.0 |
| 2006 | 250.0 | 28.0 |
| 2007 | 227.5 | 30.5 |
| 5-Year Trend (%) | -18.8% | 9.3% |

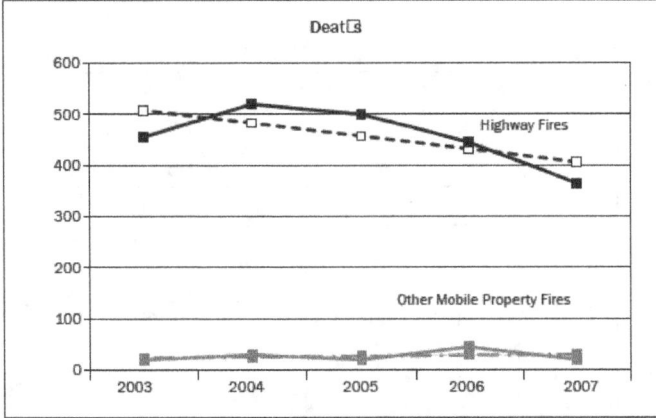

### DEATHS

| Year | Highway Vehicle Value | Other Mobile Value |
|------|------------------------|--------------------|
| 2003 | 455 | 20 |
| 2004 | 520 | 30 |
| 2005 | 500 | 20 |
| 2006 | 445 | 45 |
| 2007 | 365 | 20 |
| 5-Year Trend (%) | -20.1% | 25.0% |

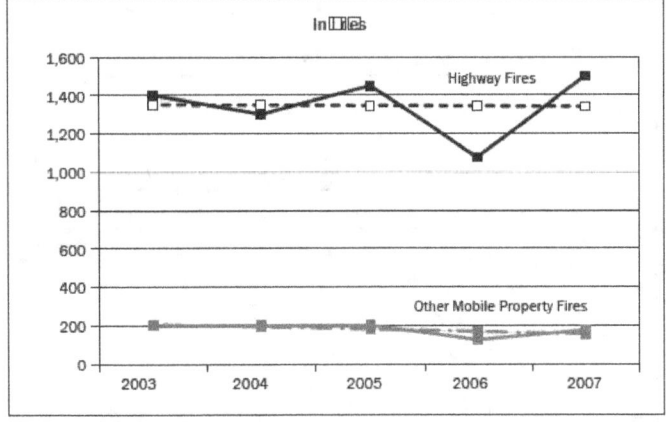

### INJURIES

| Year | Highway Vehicle Value | Other Mobile Value |
|------|------------------------|--------------------|
| 2003 | 1,400 | 200 |
| 2004 | 1,300 | 200 |
| 2005 | 1,450 | 200 |
| 2006 | 1,075 | 125 |
| 2007 | 1,500 | 175 |
| 5-Year Trend (%) | -0.7% | -24.4% |

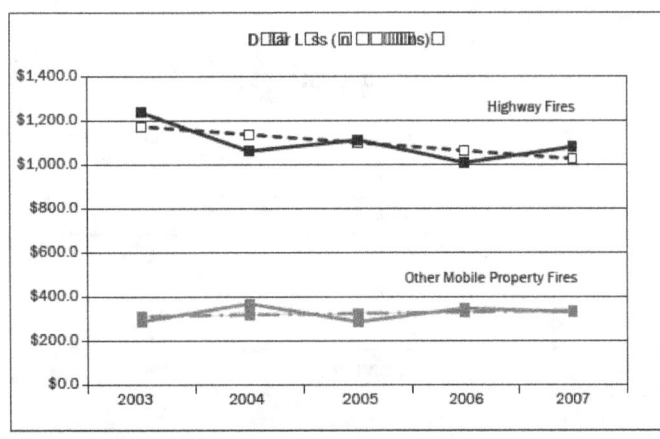

### DOLLAR LOSS ($M)*
*ADJUSTED TO 2007 DOLLARS

| Year | Highway Vehicle Value | Other Mobile Value |
|------|------------------------|--------------------|
| 2003 | $1,240.7 | $287.3 |
| 2004 | $1,063.6 | $367.7 |
| 2005 | $1,113.7 | $285.6 |
| 2006 | $1,010.0 | $346.6 |
| 2007 | $1,082.0 | $329.0 |
| 5-Year Trend (%) | -12.6% | 8.0% |

Sources: NFPA and Consumer Price Index.

# OUTSIDE AND OTHER PROPERTIES

The "Outside and Other Properties" category includes all fires that are neither buildings, other structures, nor vehicle fires. In NFIRS terminology, this includes fires where the incident types are coded as fires outside of structures, either where the burning material has a value or where the fires are confined to trees, brush, grass, or refuse. A subset of outside fires is wildland fires. Grouped in the "Other" category are fires where the incident types are not classified or are outside gas or vapor combustion incidents.

Outside and other fires comprise roughly half of all fires. This proportion, steady over the past 20 years, has continued over the 5 years (2003 to 2007). Although large in number, they accounted for only 1 percent of fire deaths in 2007, 4 percent of reported injuries, and 6 percent of reported dollar losses.[9] These numbers may not, however, reflect the true nature of the problem because of under-reporting and the difficulty in setting a price tag on outside fires. Also, many wildland fires are not reported to agencies reporting to NFIRS or to the NFPA annual survey.

## Trends

Figure 33 shows the trends in outside and other property type fires. The numbers of reported outside fires alone are enormous—averaging 645,700 over the 5-year period. The "Other" category of fires adds, on average, an additional 132,500 fires to this already large number. Over 5 years, an average of 51 deaths resulted each year from outside fires plus the miscellaneous other properties not covered elsewhere; injuries averaged 845. Although deaths have a 5-year downward trend of 24 percent, this is due primarily to the fluctuations in the small numbers of deaths; injuries also show a downward trend of 23 percent. Dollar loss for outside properties increased an enormous 1,360 percent, largely due to three large loss incidents which totaled 525 million dollars in damage.[10] Without those fires, dollar loss from outside fires still increased 43 percent over the 5 years.

Estimating dollar loss for these fires is difficult. In addition, part of the difference in property loss estimates is because NFPA estimates property loss only for outside fires "with value," whereas NFIRS permits property loss data collection for any fire. Which method is correct? Both are reasonable approaches, but neither may be definitive. Moreover, when there are large loss fires, such as in 2007, these fires may not necessarily be reported to NFIRS.

## Special Data Problems

Setting a value for outside fire damage is always a problem. It is difficult to assign a dollar value to grass, tree, and rubbish fires, yet the damage from these fires often requires labor beyond that of the fire department to clean up and restore the area. They also cause esthetic problems that are intangible. Some outside fires spread to structural properties and may be reported as structural fires rather than an outside fire with exposure to structures. Outside fires can have other indirect costs, such as the financial impact on agricultural communities where a fire destroys crops.

Forest fires and other wildfires to which local departments are not called will not be reported to NFIRS if the State or Federal agency with principal authority for fighting the fire does not participate in NFIRS. To more fully analyze outside fires, NFIRS data need to be complemented with data from these other agencies.

---

[9] Karter, Jr., Michael J., *Fire Loss in the United States 2007*, NFPA, August 2008. These percentages are derived from summary data presented in this report. Dollar loss percentages do not include the losses from the 2007 California Fire Storm as those losses were not ascribed to property types.
[10] Karter, Jr., Michael J., *Fire Loss in the United States 2007*, NFPA, August 2008.

## Figure 33. Trends in Outside and Other Property Type Fires and Fire Losses

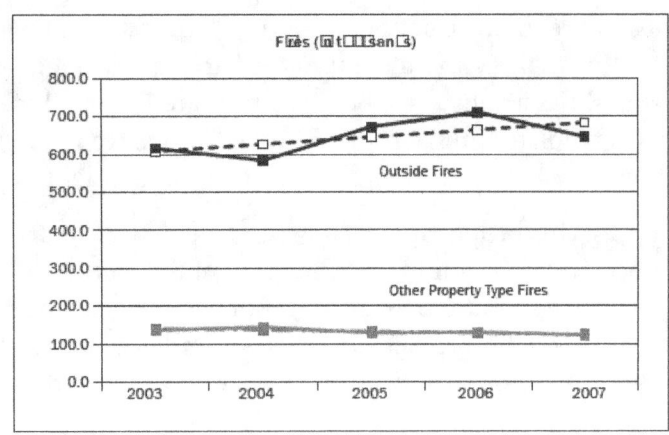

### FIRES (THOUSANDS)

| Year | Outside Value | Other Value |
|---|---|---|
| 2003 | 616.5 | 136.5 |
| 2004 | 583.0 | 144.5 |
| 2005 | 672.5 | 128.5 |
| 2006 | 710.0 | 130.5 |
| 2007 | 646.5 | 122.5 |
| 5-Year Trend (%) | 12.3% | -11.9% |

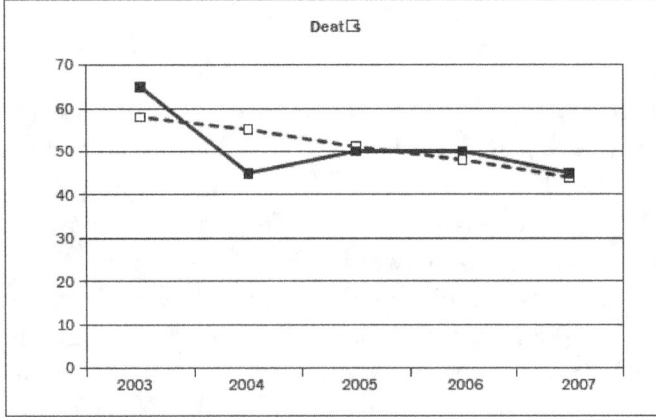

### DEATHS

| Year | Outside and Other Value |
|---|---|
| 2003 | 65 |
| 2004 | 45 |
| 2005 | 50 |
| 2006 | 50 |
| 2007 | 45 |
| 5-Year Trend (%) | -24.1% |

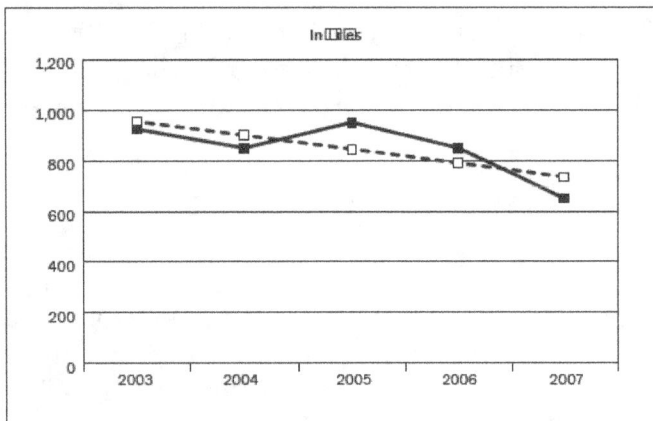

### INJURIES

| Year | Outside and Other Value |
|---|---|
| 2003 | 925 |
| 2004 | 850 |
| 2005 | 950 |
| 2006 | 850 |
| 2007 | 650 |
| 5-Year Trend (%) | -23.0% |

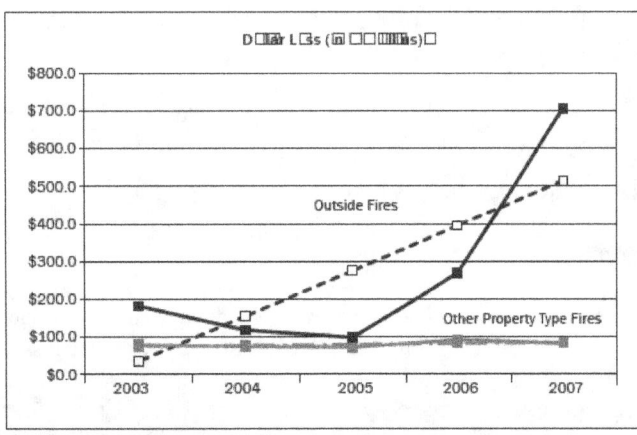

### DOLLAR LOSS ($M)*
*ADJUSTED TO 2007 DOLLARS

| Year | Outside Value | Other Value |
|---|---|---|
| 2003 | $182.6 | $80.0 |
| 2004 | $118.5 | $74.6 |
| 2005 | $98.7 | $72.2 |
| 2006 | $269.5 | $92.6 |
| 2007 | $707.0 | $83.0 |
| 5-Year Trend (%) | 1359.8% | 12.6% |

Sources:   NFPA and Consumer Price Index.

# Appendix A

# Differences Between NFPA and NFIRS Estimates

The National Fire Incident Reporting System (NFIRS) collects fire incident data from an average of 18,960 fire departments each year. The National Fire Protection Association's (NFPA) annual survey of fire departments[1] collects data from approximately 3,000 fire departments. NFIRS is not a sample; it is a very large set of fire incidents—estimated to be well over half of reported fires. The NFPA survey is based on a statistical sample. These two data sets often yield dramatically different fire rates. The NFPA survey collects tallied totals, whereas NFIRS collects individual incident reports. During the period examined, the proportion of native NFIRS 5.0 fire data rose from 81 percent of all NFIRS fire incidents collected in 2003 to 98 percent of all NFIRS fire incidents in 2007. It is not surprising, therefore, that there are differences between the NFPA annual survey results and the NFIRS results. In the years examined (2003 to 2007), the common thread is the increase in the ratios of NFIRS data to the NFPA estimates as more version 5.0 data are collected. In general, the deaths reported to NFIRS represent a smaller fraction of the NFPA national estimate of deaths than the NFIRS number of fires is of the NFPA estimate of fires. Estimates of dollar loss are notoriously inexact; it is not surprising that the NFIRS dollar loss changes from year to year with respect to NFPA totals. (Figure A-1).

---

1 "Fire Loss in the United States," NFPA Journal, generally the September/October issue each year.

## Figure A-1. Ratio of Raw NFIRS Sample to NFPA National Estimates

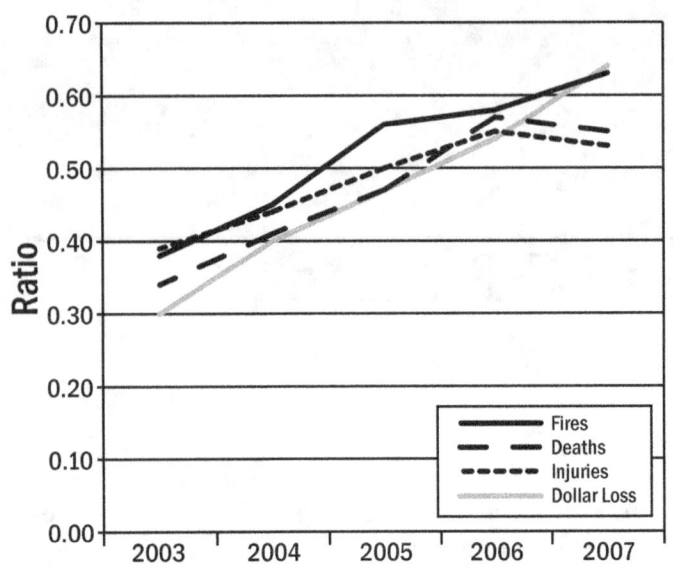

| Year | Fires | Deaths | Injuries | Dollar Loss |
|---|---|---|---|---|
| 2003 | 0.38 | 0.34 | 0.39 | 0.30 |
| 2004 | 0.45 | 0.41 | 0.44 | 0.40 |
| 2005 | 0.56 | 0.47 | 0.50 | 0.47 |
| 2006 | 0.58 | 0.57 | 0.55 | 0.54 |
| 2007 | 0.63 | 0.55 | 0.53 | 0.64 |

Note:      2003 Dollar Loss excludes the one-time large loss of an estimated $2.04B associated with 2003 Southern California Wildfires. The 2007 Dollar Loss excludes the one-time large loss of an estimated $1.8B associated with the 2007 California Fire Storm. These losses do not have associated property uses.

Sources:    NFPA and NFIRS.

Looking at the problem from a different perspective, Figure A-2 shows the number of deaths per thousand fires, injuries per thousand fires, and dollar loss per fire from NFIRS and NFPA from 2003 to 2007. In general, deaths and injuries per thousand fires and dollar loss per fire are lower for NFIRS than for NFPA. This difference may be the result of more low-loss fires being reported to NFIRS as a result of the abbreviated reporting option for these fires.

With the exception of 2006, between 2003 and 2007 NFIRS has, on average, a difference of 13 percent fewer fire deaths per thousand fires than the NFPA survey data. In 2006, NFIRS and NFPA show similar deaths per thousand fires (3 percent difference).

Injuries per thousand fires were quite close between the two data sets in 2003, 2004 and 2006 (an average difference of 2 percent, 2 percent, and 5 percent, respectively), but revealed a much greater disparity in 2005 (11 percent) and 2007 (16 percent).

Despite the abbreviated reporting for low-loss fires, the gap between NFIRS and NFPA dollar loss per fire data consistently decreased over the 5 year period. By 2007, more dollar loss was reported to NFIRS per fire than that reflected in the NFPA survey data.

## Figure A-2. NFIRS versus NFPA Survey: Losses per Fire

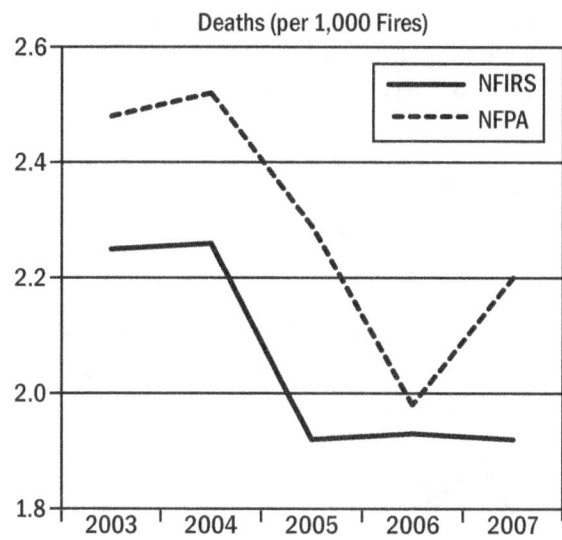

Deaths (per 1,000 Fires)

| DEATHS (per 1,000 Fires) | | |
|---|---|---|
| Year | NFIRS | NFPA |
| 2003 | 2.25 | 2.48 |
| 2004 | 2.26 | 2.52 |
| 2005 | 1.92 | 2.29 |
| 2006 | 1.93 | 1.98 |
| 2007 | 1.92 | 2.20 |

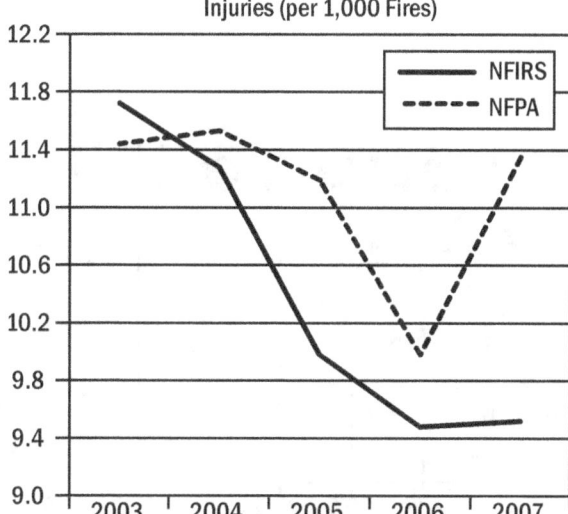

Injuries (per 1,000 Fires)

| INJURIES (per 1,000 Fires) | | |
|---|---|---|
| Year | NFIRS | NFPA |
| 2003 | 11.72 | 11.44 |
| 2004 | 11.28 | 11.53 |
| 2005 | 9.98 | 11.19 |
| 2006 | 9.48 | 9.98 |
| 2007 | 9.52 | 11.35 |

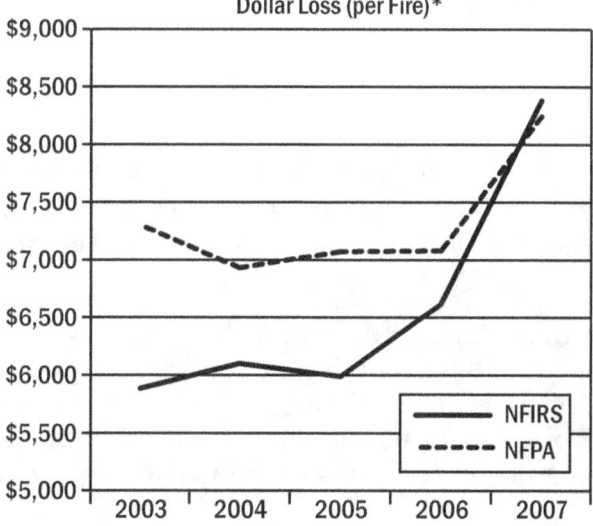

Dollar Loss (per Fire)*

| DOLLAR LOSS (per Fire)* | | |
|---|---|---|
| *Adjusted to 2007 dollars | | |
| Year | NFIRS | NFPA |
| 2003 | 5,886 | 7,302 |
| 2004 | 6,101 | 6,933 |
| 2005 | 5,990 | 7,072 |
| 2006 | 6,621 | 7,080 |
| 2007 | 8,381 | 8,243 |

Note: 2003 Dollar Loss excludes the one-time large loss of an estimated $2.04B associated with 2003 Southern California Wildfires. The 2007 Dollar Loss excludes the one-time large loss of an estimated $1.8B associated with the 2007 California Fire Storm. These losses do not have associated property uses.

Sources: NFPA, NFIRS, and Consumer Price Index.

Other minor differences appear when reviewing losses by property type as shown in Figure A-3. Specifically, the distributions of fires across property types between NFIRS and NFPA are quite similar, which is reassuring. Over the 5-year period, the proportions of structure fires (both residential and nonresidential) and outside and other fires are slightly higher in the NFIRS sample while the proportion of vehicle fires is slightly more represented in the NFPA estimate. Regardless of the specifics, the distributions are reasonably comparable.

**Figure A-3. Comparison of NFIRS Data with NFPA Estimates by General Property Type (5-year average, 2003-2007)**

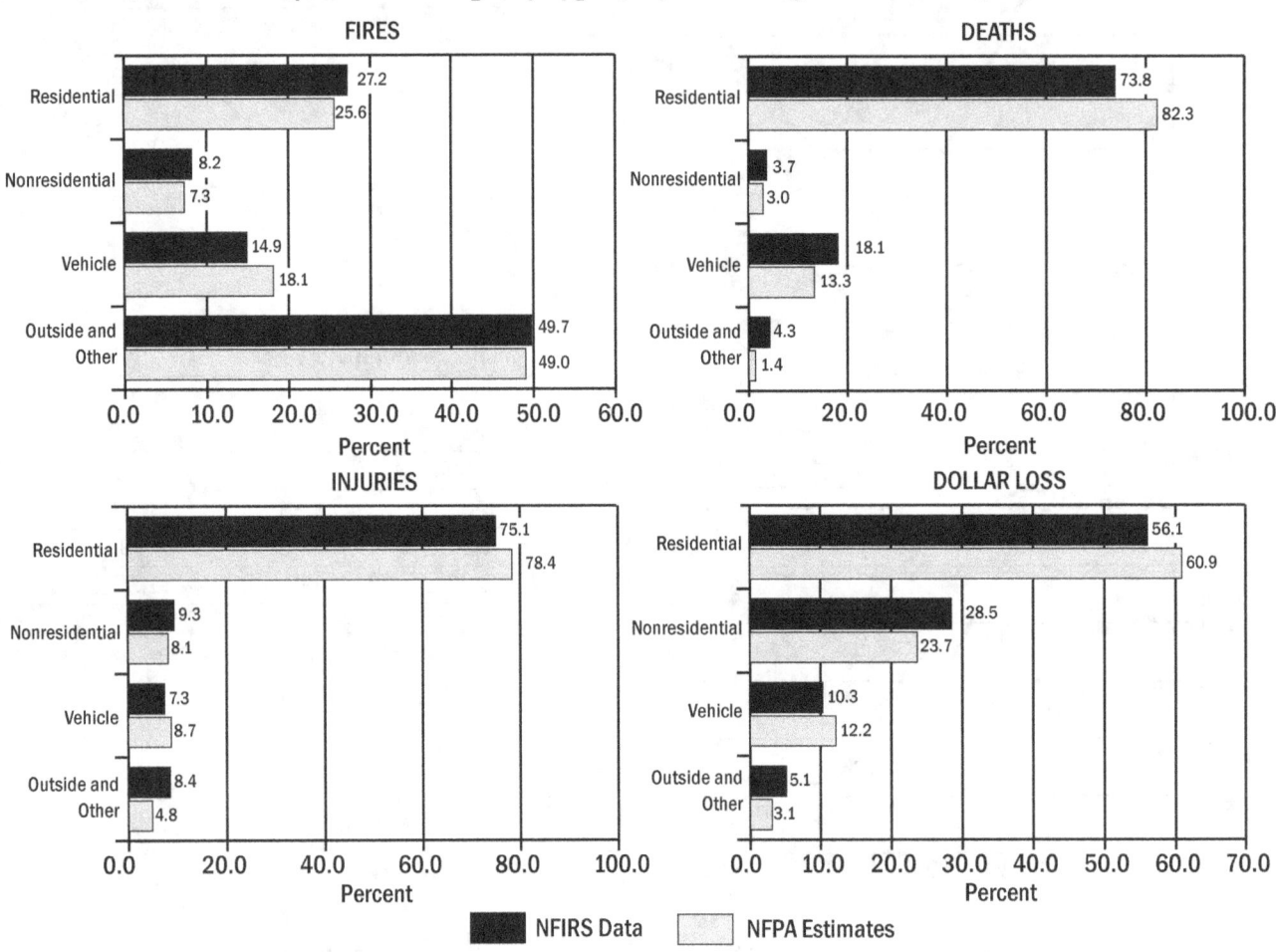

Sources:  NFPA and NFIRS.

The deaths, injuries, and dollar losses that result from these fires are consistently more heavily represented in residential structures in the NFPA estimates. For the other major property categories (except vehicular fire injuries and dollar loss), the NFPA percentages of losses are consistently less than those resulting from NFIRS data.

One of the more important consequences of these distributions is in the creation of estimates of the various parts of the U.S. fire problem. As NFPA only presents estimates of the major property types and as the distribution of these property types is not congruent to that of NFIRS, the challenge becomes how to best estimate subsets of the U.S. fire problem. For example, in Chapter 3, the number of residential building fires in 2007 can be estimated by taking the relative proportion of NFIRS residential building fires to NFIRS residential structure fires and applying it to the NFPA estimate for residential structure fires. This method results in:

The national estimate for residential building fires (rounded to the nearest 100) =

$$414{,}000 \text{ NFPA residential structure fires} \times \frac{245{,}543 \text{ NFIRS residential building fires}}{260{,}471 \text{ NFIRS residential structure fires}} = 390{,}300$$

Alternatively, the national estimate may be computed by taking the relative proportion of NFIRS residential building fires to all NFIRS fires and applying it to the NFPA estimate for all fires. This method results in:

The national estimate for residential building fires (rounded to the nearest 100) =

$$1{,}557{,}500 \text{ NFPA total fires} \times \frac{245{,}543 \text{ NFIRS residential building fires}}{978{,}796 \text{ NFIRS total fires}} = 390{,}700$$

These estimates are quite close for fires. When applied to deaths, however, there can be considerable divergence.

The national estimate for residential building fire deaths (rounded to the nearest 5) =

$$2{,}895 \text{ NFPA residential structure fire deaths} \times \frac{1{,}359 \text{ NFIRS residential building fire deaths}}{1{,}421 \text{ NFIRS residential structure fire deaths}} = 2{,}770$$

The national estimate for residential building fire deaths (rounded to the nearest 5) =

$$3{,}430 \text{ NFPA total fire deaths} \times \frac{1{,}359 \text{ NFIRS residential building fire deaths}}{1{,}881 \text{ NFIRS total fire deaths}} = 2{,}480$$

The reasons for these differences in distributions between NFPA and NFIRS are not known. It may be that some departments reporting summary data to NFPA inadvertently undercount their casualties and losses when reporting on the NFPA survey forms. Another possibility is that there are data entry errors in NFIRS, with larger numbers of deaths, injuries, and dollar loss per incident record creeping into the database despite edit checks at State and Federal levels. (It appears that at least some of the dollar loss difference is due to this.)

A third possibility for the differences is that with the introduction of abbreviated reporting of small, no-loss confined fires in NFIRS, the NFPA sample of these fires is not adequately represented. It is known that, prior to abbreviated NFIRS reporting, some departments did not fill out NFIRS forms for minor fires such as food on stoves or chimney fires. It is not clear whether these fires were or were not included in the department's report to NFPA and whether this reporting has changed. Also unknown is the actual extent of this problem.

A fourth possibility is that some jurisdictions use NFIRS as a tracking system for fire casualty information without providing the related incident data or vice versa. We know that this possibility does indeed occur from time to time in NFIRS. Again, we are unsure of how these incidents and their corresponding losses are reported to NFPA.

Lastly, it could be that techniques used to generate the NFPA estimates unintentionally favor residential buildings or that NFIRS, because it is a voluntary system and not a true statistical sample, may have a bias that results in fewer residential losses.

Resolving the differences between the two major sources of fire statistics in the United States is important to prevent confusion among users of the data.

# Appendix B

**National Fire Data, 10-Year Trends 1998 – 2007**

## Figure B-1. Fires and Fire Losses, 1998–2007

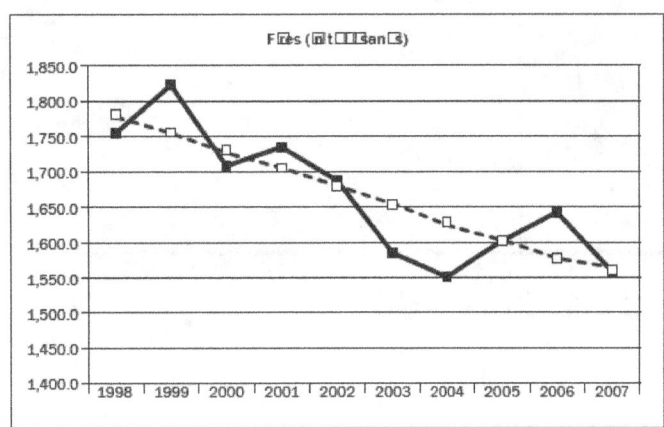

### FIRES (THOUSANDS)

| Year | Value |
|---|---|
| 1998 | 1,755.5 |
| 1999 | 1,823.0 |
| 2000 | 1,708.0 |
| 2001 | 1,734.5 |
| 2002 | 1,687.5 |
| 2003 | 1,584.5 |
| 2004 | 1,550.5 |
| 2005 | 1,602.0 |
| 2006 | 1,642.5 |
| 2007 | 1,557.5 |
| **10-Year Trend (%)** | **-13.0%** |

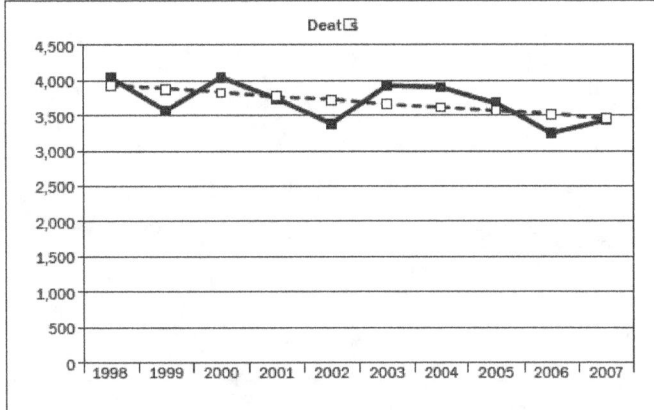

### DEATHS

| Year | Value |
|---|---|
| 1998 | 4,035 |
| 1999 | 3,570 |
| 2000 | 4,045 |
| 2001 | 3,745 |
| 2002 | 3,380 |
| 2003 | 3,925 |
| 2004 | 3,900 |
| 2005 | 3,675 |
| 2006 | 3,245 |
| 2007 | 3,430 |
| **10-Year Trend (%)** | **-11.9%** |

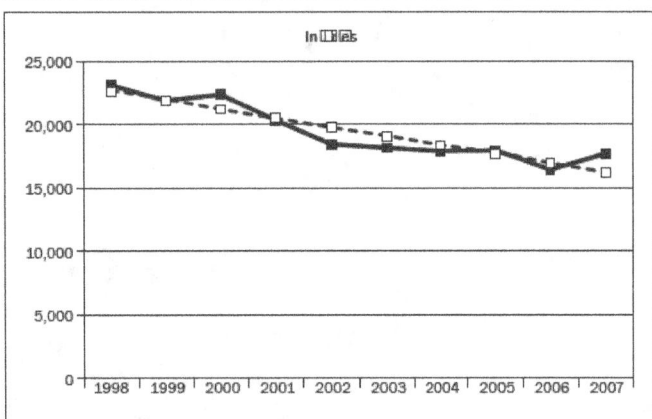

### INJURIES

| Year | Value |
|---|---|
| 1998 | 23,100 |
| 1999 | 21,875 |
| 2000 | 22,350 |
| 2001 | 20,300 |
| 2002 | 18,425 |
| 2003 | 18,125 |
| 2004 | 17,875 |
| 2005 | 17,925 |
| 2006 | 16,400 |
| 2007 | 17,675 |
| **10-Year Trend (%)** | **-28.2%** |

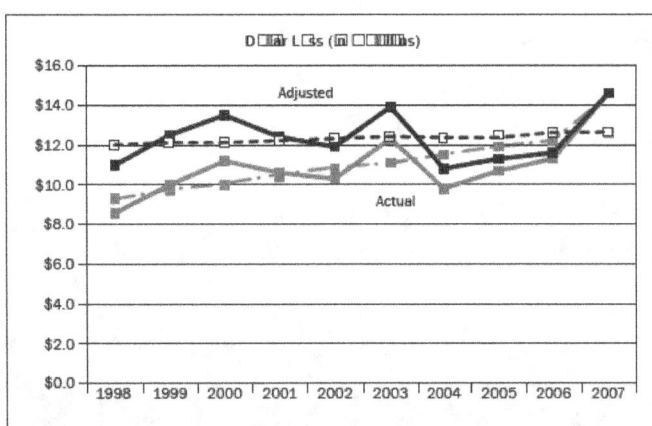

### DOLLAR LOSS ($B)

| Year | Actual Value | Adjusted to 2007 Dollars Value |
|---|---|---|
| 1998 | $8.6 | $11.0 |
| 1999 | $10.0 | $12.5 |
| 2000 | $11.2 | $13.5 |
| 2001 | $10.6 | $12.4 |
| 2002 | $10.3 | $11.9 |
| 2003 | $12.3 | $13.9 |
| 2004 | $9.8 | $10.8 |
| 2005 | $10.7 | $11.3 |
| 2006 | $11.3 | $11.6 |
| 2007 | $14.6 | $14.6 |
| **10-Year Trend (%)** | **35.1%** | **6.0%** |

Sources:   NFPA and Consumer Price Index.

## Figure B-2. Fire Loss Rates, 1998–2007

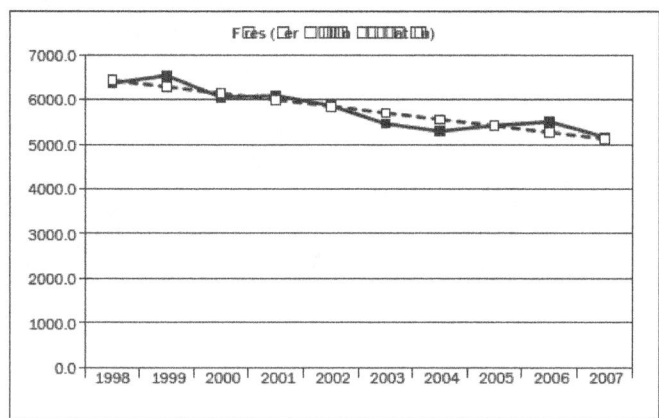

### FIRES PER MILLION POPULATION

| Year | Value |
|---|---|
| 1998 | 6,363.9 |
| 1999 | 6,533.1 |
| 2000 | 6,053.0 |
| 2001 | 6,085.1 |
| 2002 | 5,864.9 |
| 2003 | 5,459.8 |
| 2004 | 5,293.8 |
| 2005 | 5,420.2 |
| 2006 | 5,505.0 |
| 2007 | 5,169.4 |
| **10-Year Trend (%)** | **-20.3%** |

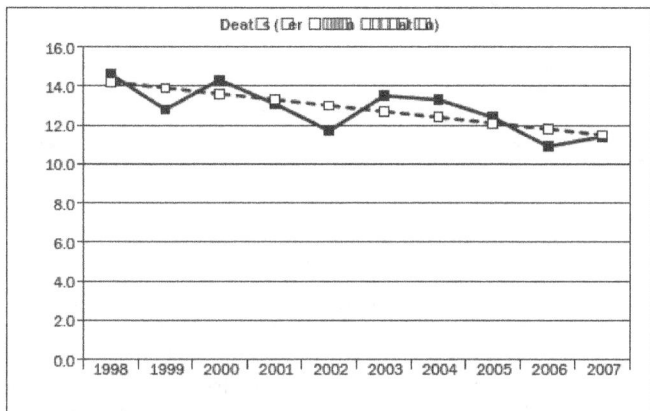

### DEATHS PER MILLION POPULATION

| Year | Value |
|---|---|
| 1998 | 14.6 |
| 1999 | 12.8 |
| 2000 | 14.3 |
| 2001 | 13.1 |
| 2002 | 11.7 |
| 2003 | 13.5 |
| 2004 | 13.3 |
| 2005 | 12.4 |
| 2006 | 10.9 |
| 2007 | 11.4 |
| **10-Year Trend (%)** | **-19.2%** |

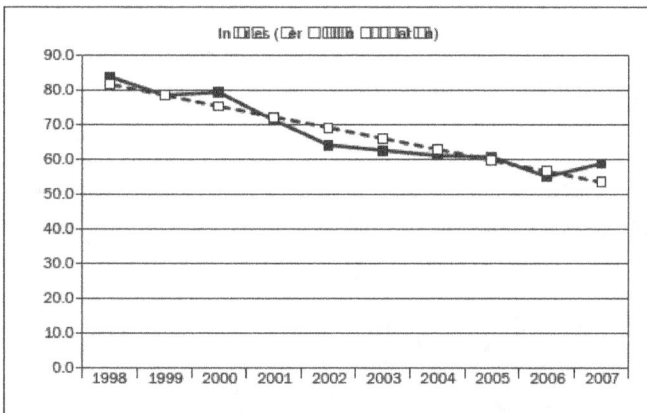

### INJURIES PER MILLION POPULATION

| Year | Value |
|---|---|
| 1998 | 83.7 |
| 1999 | 78.4 |
| 2000 | 79.2 |
| 2001 | 71.2 |
| 2002 | 64.0 |
| 2003 | 62.5 |
| 2004 | 61.0 |
| 2005 | 60.6 |
| 2006 | 55.0 |
| 2007 | 58.7 |
| **10-Year Trend (%)** | **-34.5%** |

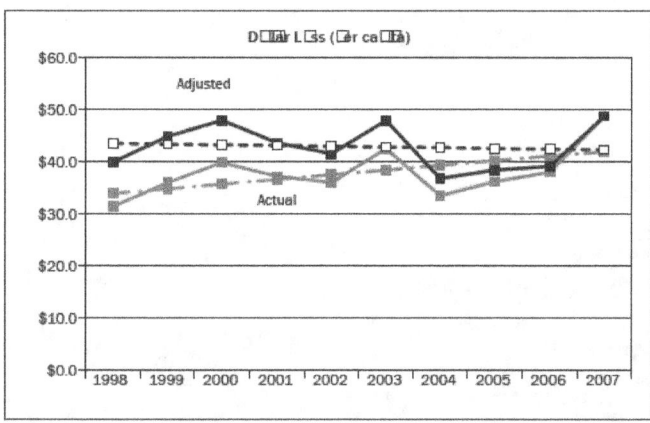

### DOLLAR LOSS PER CAPITA

| Year | Actual Value | Adjusted Value |
|---|---|---|
| 1998 | $31.3 | $39.8 |
| 1999 | $35.9 | $44.7 |
| 2000 | $39.7 | $47.8 |
| 2001 | $37.1 | $43.5 |
| 2002 | $35.9 | $41.4 |
| 2003 | $42.4 | $47.8 |
| 2004 | $33.4 | $36.7 |
| 2005 | $36.1 | $38.3 |
| 2006 | $37.9 | $39.0 |
| 2007 | $48.6 | $48.6 |
| **10-Year Trend (%)** | **23.7%** | **-2.8%** |

Sources: NFPA, Consumer Price Index, and U.S. Census Bureau.

# Appendix C

**Fire Data by Property Use, 10-Year Trends 1998 – 2007**

## Figure C- 1. Trends in Residential Structure Fires and Fire Losses, 1998-2007

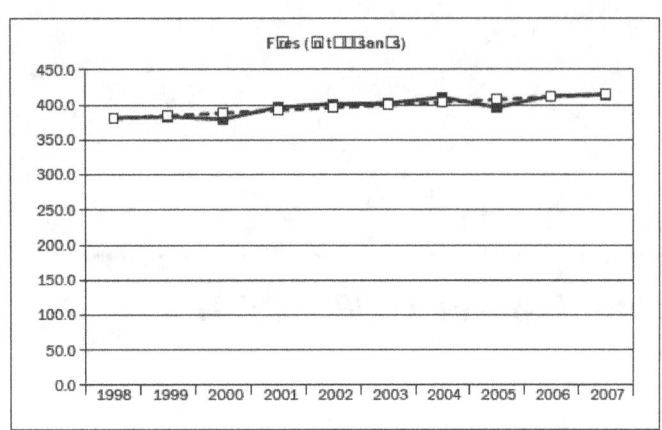

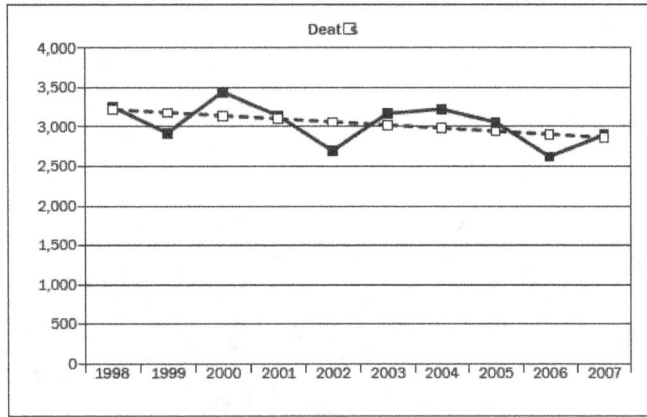

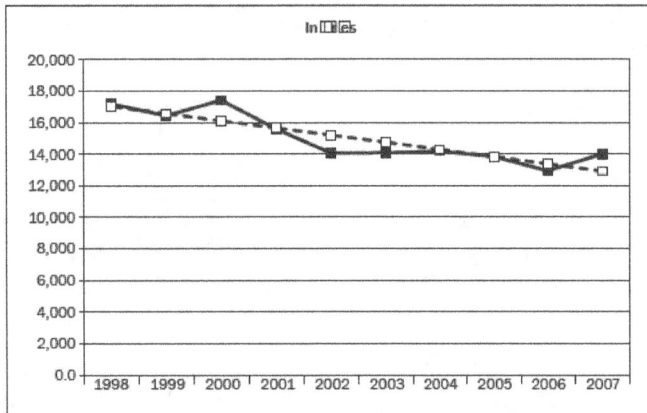

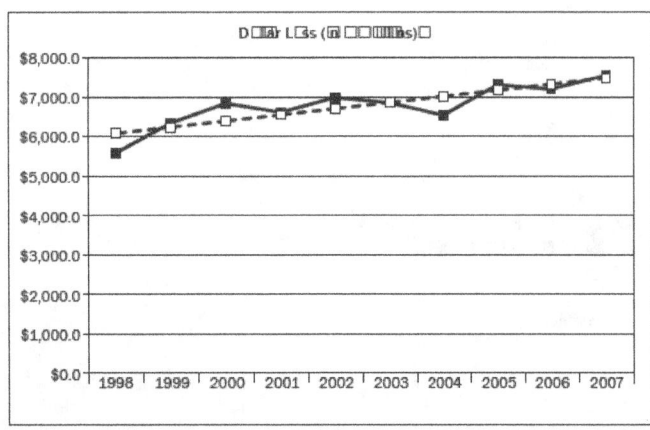

Sources: NFPA and Consumer Price Index.

### FIRES (THOUSANDS)

| Year | Value |
|---|---|
| 1998 | 381.5 |
| 1999 | 383.0 |
| 2000 | 379.5 |
| 2001 | 396.5 |
| 2002 | 401.0 |
| 2003 | 402.0 |
| 2004 | 410.5 |
| 2005 | 396.0 |
| 2006 | 412.5 |
| 2007 | 414.0 |
| **10-Year Trend (%)** | **8.9%** |

### DEATHS

| Year | Value |
|---|---|
| 1998 | 3,250 |
| 1999 | 2,920 |
| 2000 | 3,445 |
| 2001 | 3,140 |
| 2002 | 2,695 |
| 2003 | 3,165 |
| 2004 | 3,225 |
| 2005 | 3,055 |
| 2006 | 2,620 |
| 2007 | 2,895 |
| **10-Year Trend (%)** | **-11.0%** |

### INJURIES

| Year | Value |
|---|---|
| 1998 | 17,175 |
| 1999 | 16,425 |
| 2000 | 17,400 |
| 2001 | 15,575 |
| 2002 | 14,050 |
| 2003 | 14,075 |
| 2004 | 14,175 |
| 2005 | 13,825 |
| 2006 | 12,925 |
| 2007 | 14,000 |
| **10-Year Trend (%)** | **-24.1%** |

### DOLLAR LOSS ($M)*
*ADJUSTED TO 2007 DOLLARS

| Year | Value |
|---|---|
| 1998 | $5,585.5 |
| 1999 | $6,337.2 |
| 2000 | $6,831.9 |
| 2001 | $6,606.6 |
| 2002 | $6,978.6 |
| 2003 | $6,844.5 |
| 2004 | $6,528.7 |
| 2005 | $7,298.9 |
| 2006 | $7,189.1 |
| 2007 | $7,546.0 |
| **10-Year Trend (%)** | **23.0%** |

# Figure C-2. Trends in One- and Two-Family Dwelling Structure Fires and Fire Losses, 1998-2007

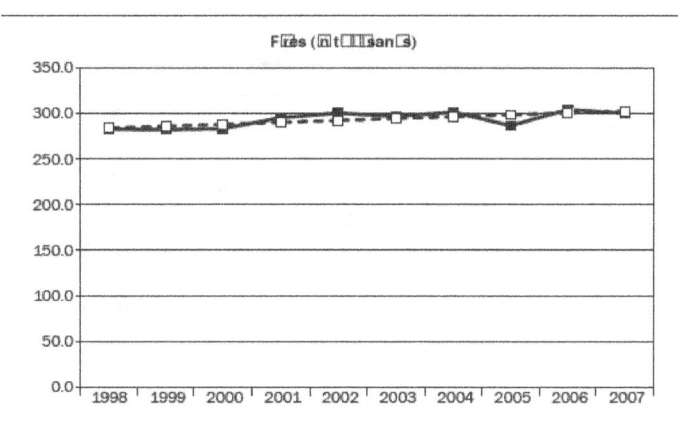

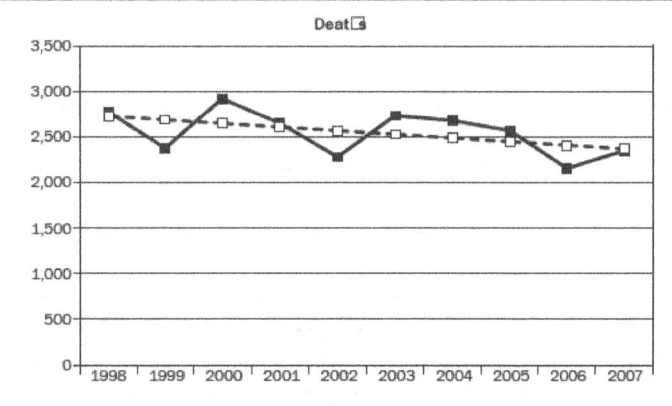

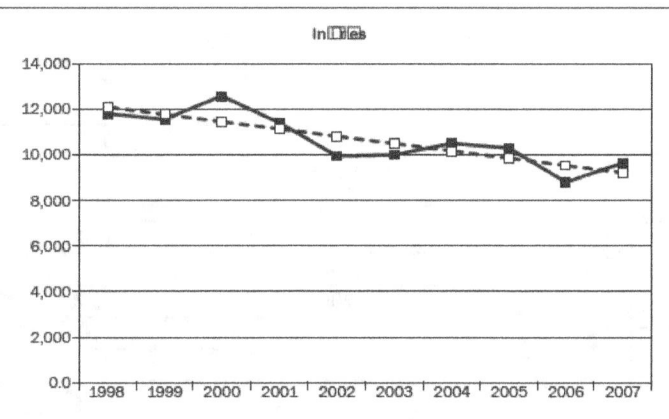

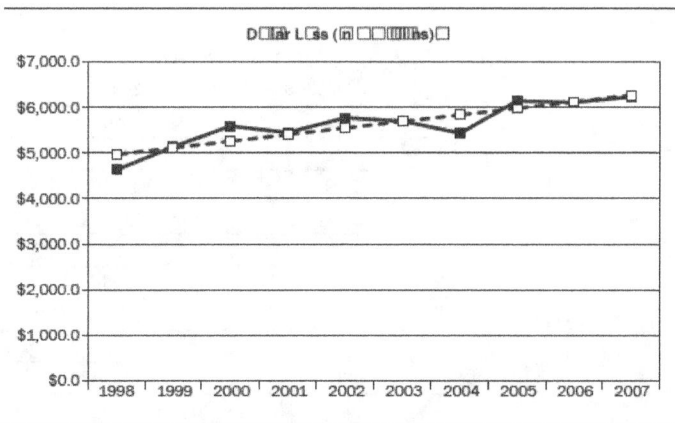

Sources: NFPA and Consumer Price Index.

**FIRES (THOUSANDS)**

| Year | Value |
|---|---|
| 1998 | 283.0 |
| 1999 | 282.5 |
| 2000 | 283.5 |
| 2001 | 295.5 |
| 2002 | 300.5 |
| 2003 | 297.0 |
| 2004 | 301.5 |
| 2005 | 287.0 |
| 2006 | 304.5 |
| 2007 | 300.5 |
| **10-Year Trend (%)** | **6.6%** |

**DEATHS**

| Year | Value |
|---|---|
| 1998 | 2,776 |
| 1999 | 2,375 |
| 2000 | 2,920 |
| 2001 | 2,650 |
| 2002 | 2,280 |
| 2003 | 2,735 |
| 2004 | 2,680 |
| 2005 | 2,570 |
| 2006 | 2,155 |
| 2007 | 2,350 |
| **10-Year Trend (%)** | **-13.2%** |

**INJURIES**

| Year | Value |
|---|---|
| 1998 | 11,800 |
| 1999 | 11,550 |
| 2000 | 12,575 |
| 2001 | 11,400 |
| 2002 | 9,950 |
| 2003 | 10,000 |
| 2004 | 10,500 |
| 2005 | 10,300 |
| 2006 | 8,800 |
| 2007 | 9,650 |
| **10-Year Trend (%)** | **-23.7%** |

**DOLLAR LOSS ($M)***

*ADJUSTED TO 2007 DOLLARS

| Year | Value |
|---|---|
| 1998 | $4,632.8 |
| 1999 | $5,131.3 |
| 2000 | $5,585.7 |
| 2001 | $5,446.4 |
| 2002 | $5,768.5 |
| 2003 | $5,692.9 |
| 2004 | $5,431.1 |
| 2005 | $6,137.5 |
| 2006 | $6,105.1 |
| 2007 | $6,225.0 |
| **10-Year Trend (%)** | **26.1%** |

## Figure C-3. Trends in Multifamily Structure Fires and Fire Losses, 1998–2007

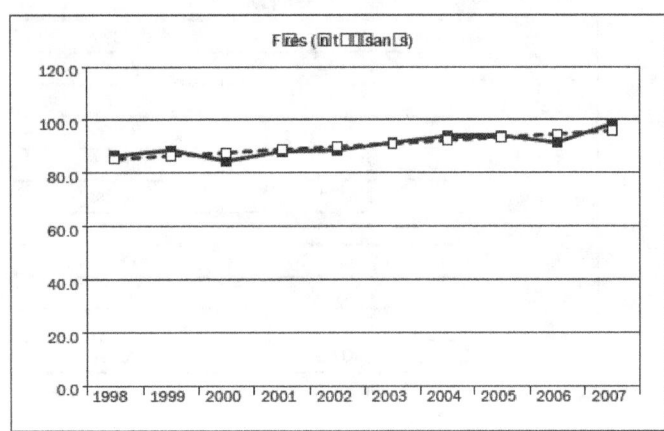

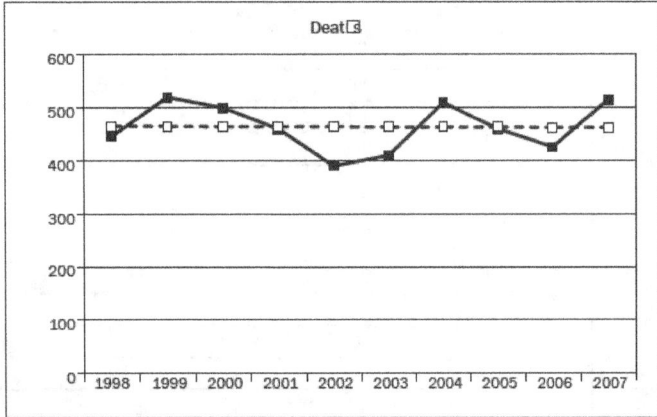

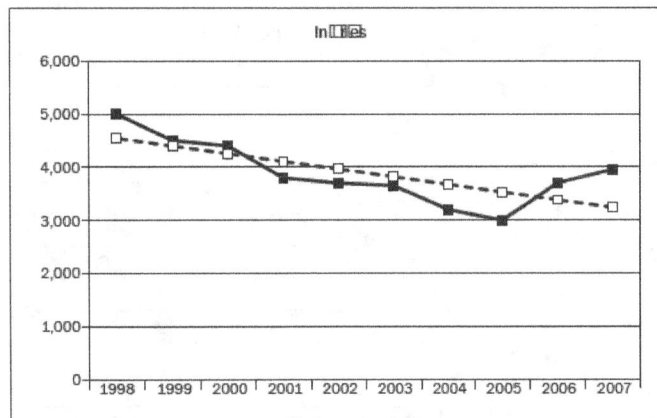

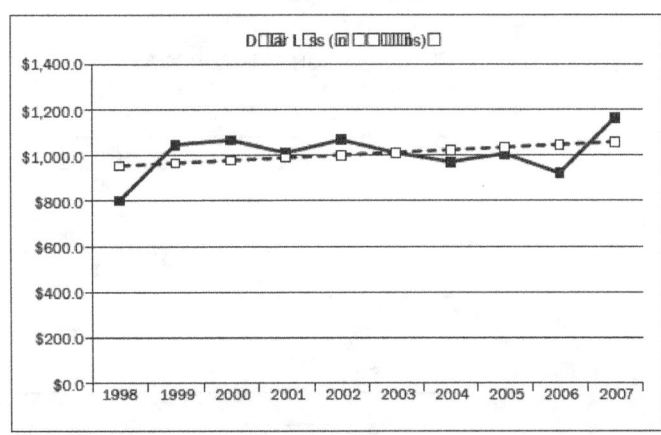

Sources: NFPA and Consumer Price Index.

### FIRES (THOUSANDS)

| Year | Value |
|------|-------|
| 1998 | 86.5 |
| 1999 | 88.5 |
| 2000 | 84.5 |
| 2001 | 88.0 |
| 2002 | 88.5 |
| 2003 | 91.5 |
| 2004 | 94.0 |
| 2005 | 94.0 |
| 2006 | 91.5 |
| 2007 | 98.5 |
| **10-Year Trend (%)** | **12.6%** |

### DEATHS

| Year | Value |
|------|-------|
| 1998 | 445 |
| 1999 | 520 |
| 2000 | 500 |
| 2001 | 460 |
| 2002 | 390 |
| 2003 | 410 |
| 2004 | 510 |
| 2005 | 460 |
| 2006 | 425 |
| 2007 | 515 |
| **10-Year Trend (%)** | **-0.8%** |

### INJURIES

| Year | Value |
|------|-------|
| 1998 | 5,000 |
| 1999 | 4,500 |
| 2000 | 4,400 |
| 2001 | 3,800 |
| 2002 | 3,700 |
| 2003 | 3,650 |
| 2004 | 3,200 |
| 2005 | 3,000 |
| 2006 | 3,700 |
| 2007 | 3,950 |
| **10-Year Trend (%)** | **-28.7%** |

### DOLLAR LOSS ($M)*
*ADJUSTED TO 2007 DOLLARS

| Year | Value |
|------|-------|
| 1998 | $802.7 |
| 1999 | $1,047.9 |
| 2000 | $1,066.8 |
| 2001 | $1,011.5 |
| 2002 | $1,067.3 |
| 2003 | $1,010.8 |
| 2004 | $971.4 |
| 2005 | $1,006.5 |
| 2006 | $921.5 |
| 2007 | $1,164.0 |
| **10-Year Trend (%)** | **10.8%** |

## Figure C-4. Trends in Other Residential Property Fires and Fire Losses, 1998-2007

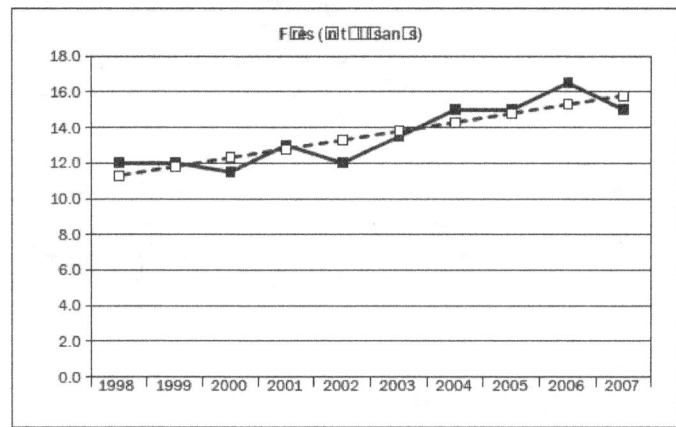

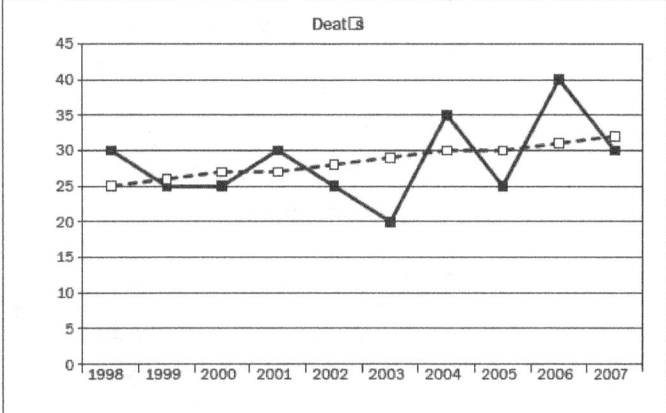

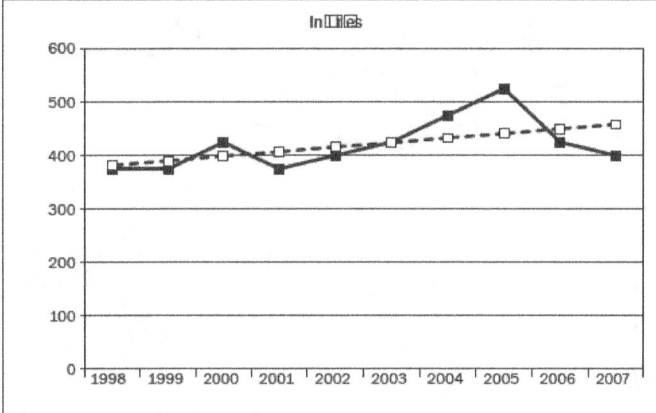

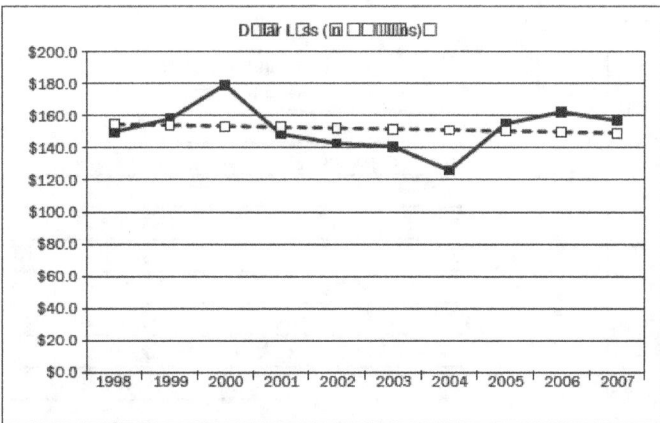

Sources: NFPA and Consumer Price Index.

### FIRES (THOUSANDS)

| Year | Value |
|---|---|
| 1998 | 12.0 |
| 1999 | 12.0 |
| 2000 | 11.5 |
| 2001 | 13.0 |
| 2002 | 12.0 |
| 2003 | 13.5 |
| 2004 | 15.0 |
| 2005 | 15.0 |
| 2006 | 16.5 |
| 2007 | 15.0 |
| **10-Year Trend (%)** | **40.4%** |

### DEATHS

| Year | Value |
|---|---|
| 1998 | 30 |
| 1999 | 25 |
| 2000 | 25 |
| 2001 | 30 |
| 2002 | 25 |
| 2003 | 20 |
| 2004 | 35 |
| 2005 | 25 |
| 2006 | 40 |
| 2007 | 30 |
| **10-Year Trend (%)** | **24.7%** |

### INJURIES

| Year | Value |
|---|---|
| 1998 | 375 |
| 1999 | 375 |
| 2000 | 425 |
| 2001 | 375 |
| 2002 | 400 |
| 2003 | 425 |
| 2004 | 475 |
| 2005 | 525 |
| 2006 | 425 |
| 2007 | 400 |
| **10-Year Trend (%)** | **20.0%** |

### DOLLAR LOSS ($M)*
*ADJUSTED TO 2007 DOLLARS

| Year | Value |
|---|---|
| 1998 | $150.1 |
| 1999 | $158.1 |
| 2000 | $179.4 |
| 2001 | $148.7 |
| 2002 | $142.9 |
| 2003 | $140.9 |
| 2004 | $126.2 |
| 2005 | $155.0 |
| 2006 | $162.5 |
| 2007 | $157.0 |
| **10-Year Trend (%)** | **-3.5** |

## Figure C-5. Trends in Nonresidential Structure Fires and Fire Losses, 1998–2007

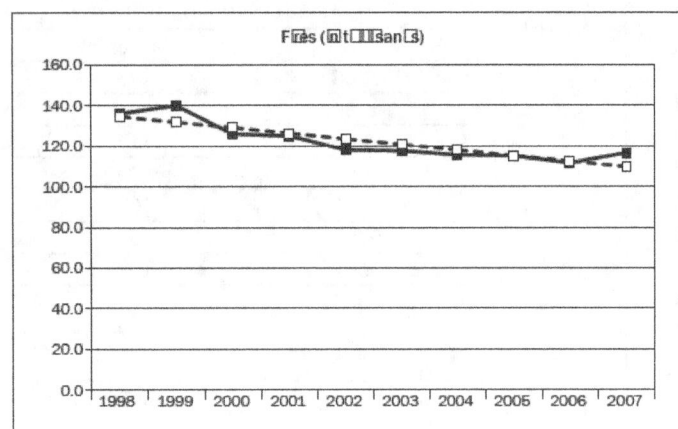

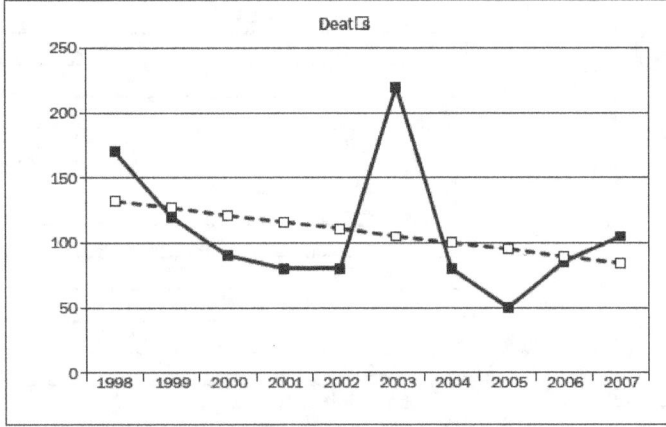

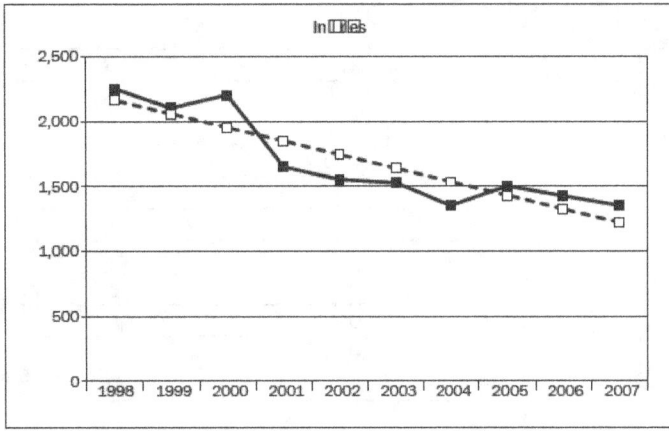

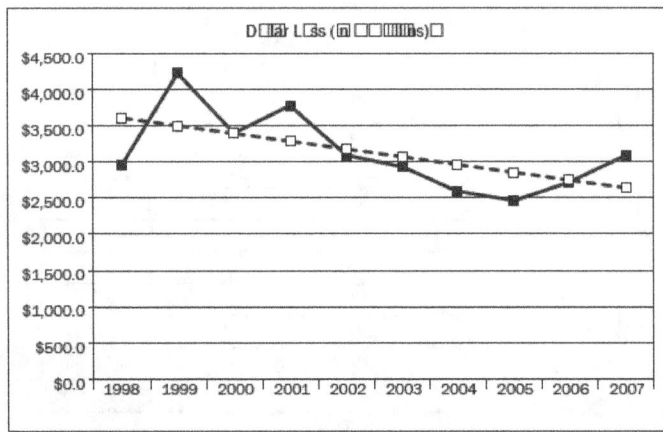

Sources: NFPA and Consumer Price Index.

### FIRES (THOUSANDS)

| Year | Value |
|---|---|
| 1998 | 136.0 |
| 1999 | 140.0 |
| 2000 | 126.0 |
| 2001 | 125.0 |
| 2002 | 118.0 |
| 2003 | 117.5 |
| 2004 | 115.5 |
| 2005 | 115.0 |
| 2006 | 111.5 |
| 2007 | 116.5 |
| **10-Year Trend (%)** | **-18.6%** |

### DEATHS

| Year | Value |
|---|---|
| 1998 | 170 |
| 1999 | 120 |
| 2000 | 90 |
| 2001 | 80 |
| 2002 | 80 |
| 2003 | 220 |
| 2004 | 80 |
| 2005 | 50 |
| 2006 | 85 |
| 2007 | 105 |
| **10-Year Trend (%)** | **-36.7%** |

### INJURIES

| Year | Value |
|---|---|
| 1998 | 2,250 |
| 1999 | 2,100 |
| 2000 | 2,200 |
| 2001 | 1,650 |
| 2002 | 1,550 |
| 2003 | 1,525 |
| 2004 | 1,350 |
| 2005 | 1,500 |
| 2006 | 1,425 |
| 2007 | 1,350 |
| **10-Year Trend (%)** | **-43.6%** |

### DOLLAR LOSS ($M)*
*ADJUSTED TO 2007 DOLLARS

| Year | Value |
|---|---|
| 1998 | $2,958.8 |
| 1999 | $4,229.0 |
| 2000 | $3,403.9 |
| 2001 | $3,782.7 |
| 2002 | $3,096.9 |
| 2003 | $2,934.3 |
| 2004 | $2,597.0 |
| 2005 | $2,460.9 |
| 2006 | $2,721.4 |
| 2007 | $3,092.0 |
| **10-Year Trend (%)** | **-26.9%** |

# Figure C-6. Trends in Mobile Property Fires and Fire Losses, 1998-2007

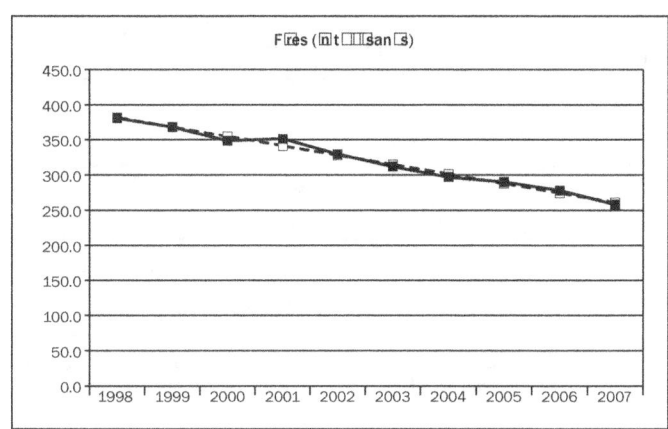

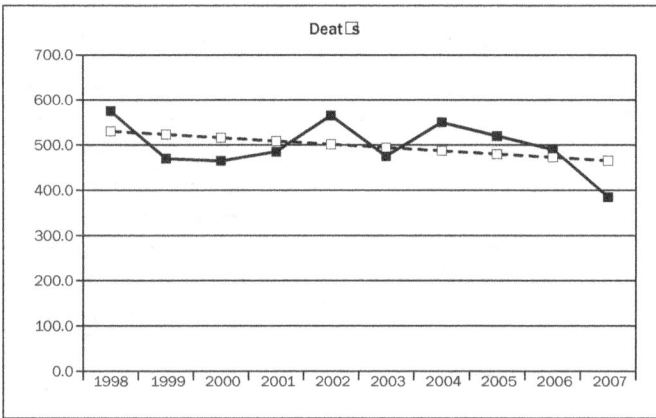

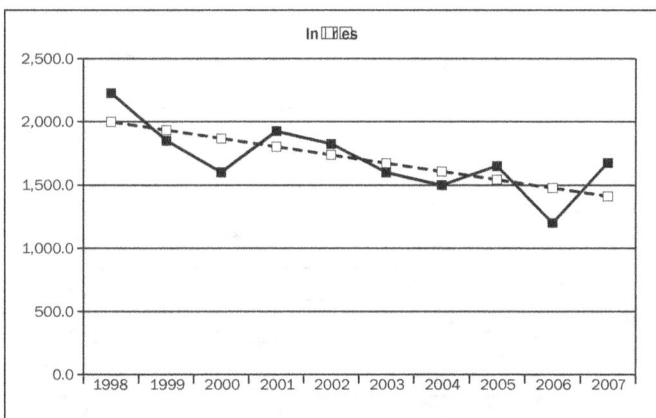

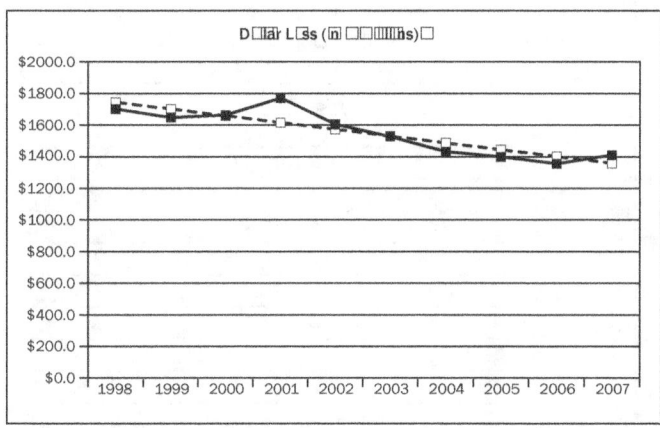

Sources: NFPA and Consumer Price Index.

### FIRES (THOUSANDS)

| Year | Value |
|---|---|
| 1998 | 381.0 |
| 1999 | 368.5 |
| 2000 | 348.5 |
| 2001 | 351.5 |
| 2002 | 329.5 |
| 2003 | 312.0 |
| 2004 | 297.0 |
| 2005 | 290.0 |
| 2006 | 278.0 |
| 2007 | 258.0 |
| **10-Year Trend (%)** | **-31.6%** |

### DEATHS

| Year | Value |
|---|---|
| 1998 | 575.0 |
| 1999 | 470.0 |
| 2000 | 465.0 |
| 2001 | 485.0 |
| 2002 | 565.0 |
| 2003 | 475.0 |
| 2004 | 550.0 |
| 2005 | 520.0 |
| 2006 | 490.0 |
| 2007 | 385.0 |
| **10-Year Trend (%)** | **-12.2%** |

### INJURIES

| Year | Value |
|---|---|
| 1998 | 2,225.0 |
| 1999 | 1,850.0 |
| 2000 | 1,600.0 |
| 2001 | 1,925.0 |
| 2002 | 1,825.0 |
| 2003 | 1,600.0 |
| 2004 | 1,500.0 |
| 2005 | 1,650.0 |
| 2006 | 1,200.0 |
| 2007 | 1,675.0 |
| **10-Year Trend (%)** | **-29.3%** |

### DOLLAR LOSS ($M)*
*ADJUSTED TO 2007 DOLLARS

| Year | Value |
|---|---|
| 1998 | $1,700.7 |
| 1999 | $1,647.8 |
| 2000 | $1,662.8 |
| 2001 | $1,770.2 |
| 2002 | $1,604.3 |
| 2003 | $1,528.0 |
| 2004 | $1,431.3 |
| 2005 | $1,399.3 |
| 2006 | $1,356.6 |
| 2007 | $1,411.0 |
| **10-Year Trend (%)** | **-22.1%** |

## Figure C-7. Trends in Highway Vehicle vs. Other Mobile Property Fires and Fire Losses, 1998–2007

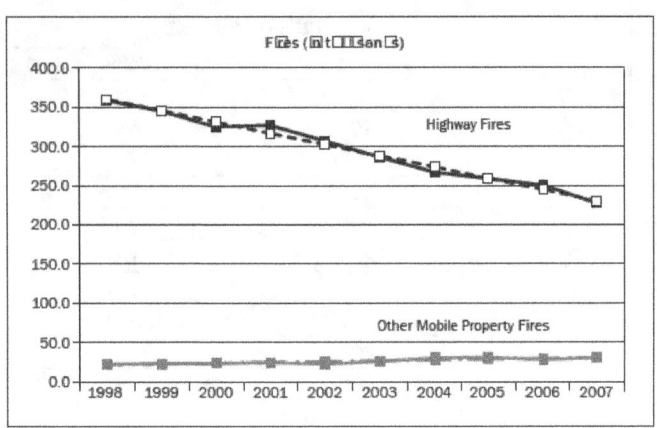

### FIRES (THOUSANDS)

| Year | Highway Vehicle | Other |
|---|---|---|
| 1998 | 358.5 | 22.5 |
| 1999 | 345.0 | 23.5 |
| 2000 | 325.0 | 23.5 |
| 2001 | 327.0 | 24.5 |
| 2002 | 307.0 | 22.5 |
| 2003 | 286.0 | 26.0 |
| 2004 | 266.5 | 30.5 |
| 2005 | 259.0 | 31.0 |
| 2006 | 250.0 | 28.0 |
| 2007 | 227.5 | 30.5 |
| 10-Year Trend (%) | -36.0% | 40.6% |

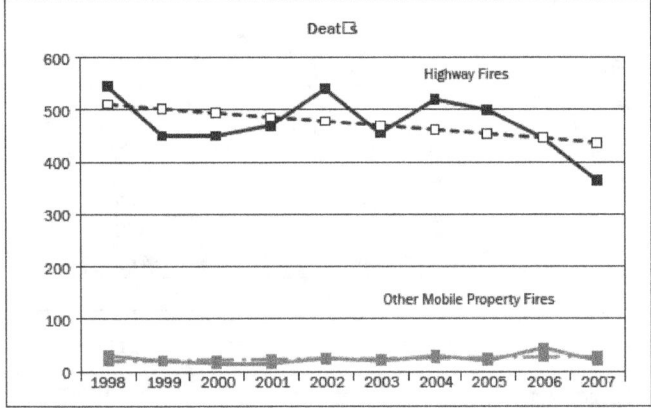

### DEATHS

| Year | Highway Vehicle | Other |
|---|---|---|
| 1998 | 545 | 30 |
| 1999 | 450 | 20 |
| 2000 | 450 | 15 |
| 2001 | 470 | 15 |
| 2002 | 540 | 25 |
| 2003 | 455 | 20 |
| 2004 | 520 | 30 |
| 2005 | 500 | 20 |
| 2006 | 445 | 45 |
| 2007 | 365 | 20 |
| 10-Year Trend (%) | -14.3% | 41.1% |

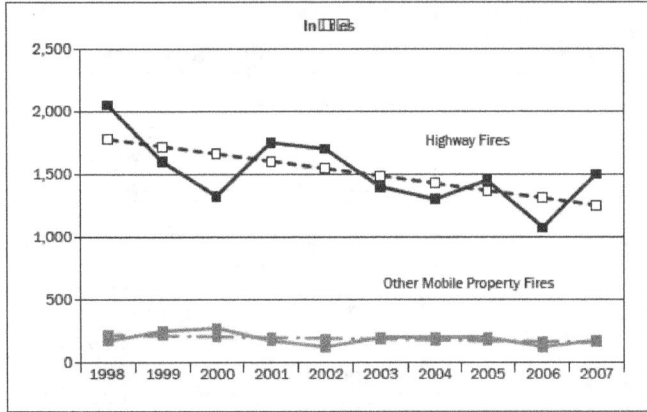

### INJURIES

| Year | Highway Vehicle | Other |
|---|---|---|
| 1998 | 2,050 | 175 |
| 1999 | 1,600 | 250 |
| 2000 | 1,325 | 275 |
| 2001 | 1,750 | 175 |
| 2002 | 1,700 | 125 |
| 2003 | 1,400 | 200 |
| 2004 | 1,300 | 200 |
| 2005 | 1,450 | 200 |
| 2006 | 1,075 | 125 |
| 2007 | 1,500 | 175 |
| 10-Year Trend (%) | -29.6% | -27.3% |

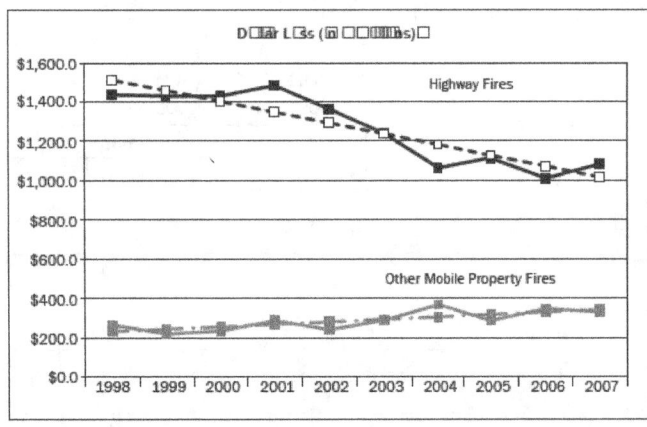

Sources:  NFPA and Consumer Price Index.

### DOLLAR LOSS ($M)*
*ADJUSTED TO 2007 DOLLARS

| Year | Highway Vehicle | Other |
|---|---|---|
| 1998 | $1,436.1 | $264.6 |
| 1999 | $1,430.0 | $217.8 |
| 2000 | $1,429.2 | $233.6 |
| 2001 | $1,483.4 | $286.8 |
| 2002 | $1,364.6 | $239.7 |
| 2003 | $1,240.7 | $287.3 |
| 2004 | $1,063.6 | $367.7 |
| 2005 | $1,113.7 | $285.6 |
| 2006 | $1,010.0 | $346.6 |
| 2007 | $1,082.0 | $329.0 |
| 10-Year Trend (%) | -32.8% | 48.1% |

## Figure C-8. Trends in Outside vs. Other Property Type Fires and Fire Losses, 1998-2007

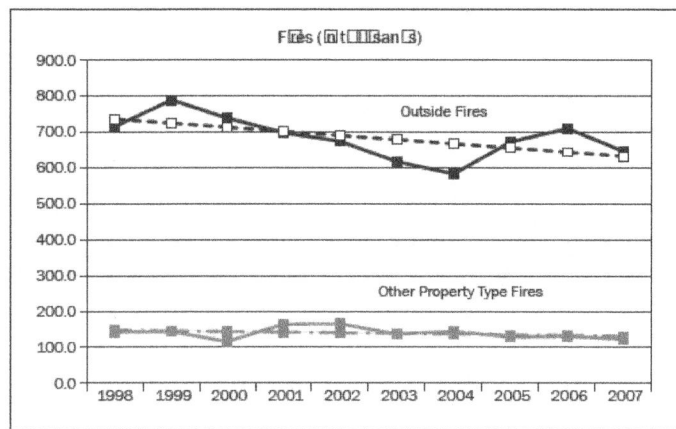

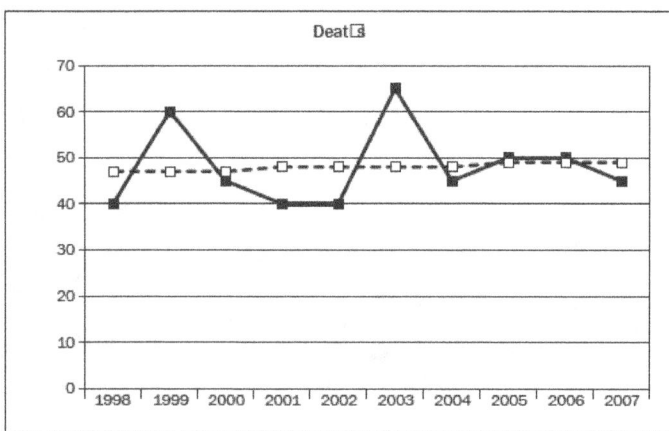

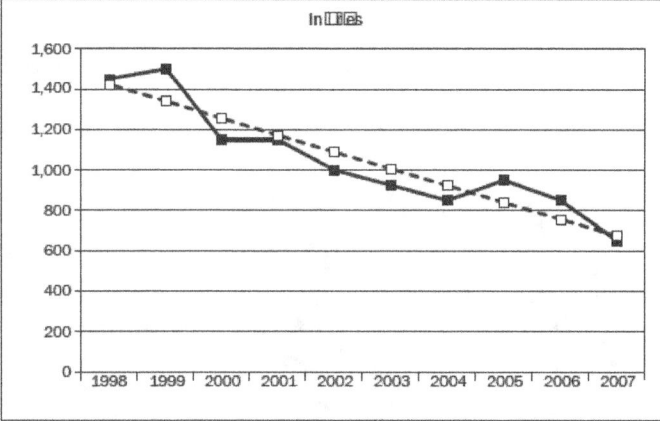

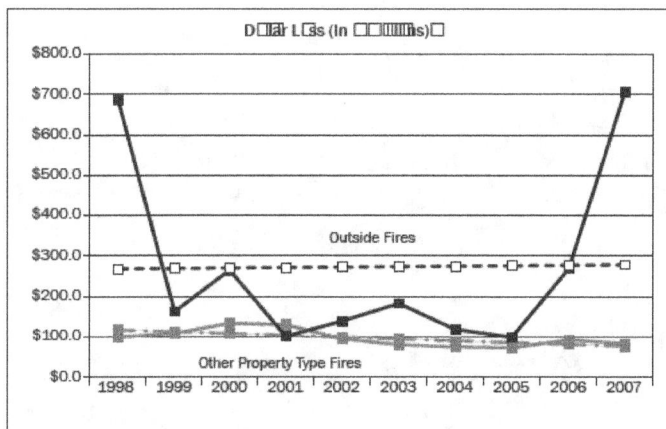

Sources: NFPA and Consumer Price Index.

### FIRES (THOUSANDS)

| Year | Outside | Other |
|------|---------|-------|
| 1998 | 715.0 | 142.0 |
| 1999 | 788.5 | 143.0 |
| 2000 | 738.5 | 115.5 |
| 2001 | 697.5 | 164.0 |
| 2002 | 674.0 | 165.0 |
| 2003 | 616.5 | 136.5 |
| 2004 | 583.0 | 144.5 |
| 2005 | 672.5 | 128.5 |
| 2006 | 710.0 | 130.5 |
| 2007 | 646.5 | 122.5 |
| **10-Year Trend (%)** | **-14.1%** | **-10.6%** |

### DEATHS

| Year | Outside and Other Value |
|------|--------------------------|
| 1998 | 40 |
| 1999 | 60 |
| 2000 | 45 |
| 2001 | 40 |
| 2002 | 40 |
| 2003 | 65 |
| 2004 | 45 |
| 2005 | 50 |
| 2006 | 50 |
| 2007 | 45 |
| **10-Year Trend (%)** | **4.7%** |

### INJURIES

| Year | Outside and Other Value |
|------|--------------------------|
| 1998 | 1,450 |
| 1999 | 1,500 |
| 2000 | 1,150 |
| 2001 | 1,150 |
| 2002 | 1,000 |
| 2003 | 925 |
| 2004 | 850 |
| 2005 | 950 |
| 2006 | 850 |
| 2007 | 650 |
| **10-Year Trend (%)** | **-52.7%** |

### DOLLAR LOSS ($M)*
*ADJUSTED TO 2007 DOLLARS

| Year | Outside | Other |
|------|---------|-------|
| 1998 | $686.9 | $99.2 |
| 1999 | $162.7 | $108.3 |
| 2000 | $265.0 | $133.7 |
| 2001 | $100.7 | $130.0 |
| 2002 | $139.5 | $94.5 |
| 2003 | $182.6 | $80.0 |
| 2004 | $118.5 | $74.6 |
| 2005 | $98.7 | $72.2 |
| 2006 | $269.5 | $92.6 |
| 2007 | $707.0 | $83.0 |
| **10-Year Trend (%)** | **3.9%** | **-34.6%** |

# Index